Human Survivability Studies

Human Survivability Studies

A New Paradigm for Solving Global Issues

Edited by

KAWAI Shuichi
FUJITA Masakatsu and
KAWAI Eriko

Kyoto University Press

TRANS PACIFIC PRESS

This English edition published in 2018 jointly by:

Kyoto University Press
69 Yoshida Konoe-cho
Sakyo-ku, Kyoto 606-8315, Japan
Telephone: +81-75-761-6182
Fax: +81-75-761-6190
Email: sales@kyoto-up.or.jp
Web: http://www.kyoto-up.or.jp

Trans Pacific Press
PO Box 164, Balwyn North
Victoria 3104, Australia
Telephone: +61-(0)3-9859-1112
Fax: +61-(0)3-8611-7989
Email: tpp.mail@gmail.com
Web: http://www.transpacificpress.com

© Kyoto University Press and Trans Pacific Press 2018.

Designed and set by Sarah Tuke, Melbourne, Australia.

Distributors

Australia and New Zealand
James Bennett Pty Ltd
Locked Bag 537
Frenchs Forest NSW 2086
Australia
Telephone: +61-(0)2-8988-5000
Fax: +61-(0)2-8988-5031
Email: info@bennett.com.au
Web: www.bennett.com.au

USA and Canada
International Specialized Book
Services (ISBS)
920 NE 58th Avenue, Suite 300
Portland, Oregon 97213-3786
USA
Telephone: 1-800-944-6190
Fax: 1-503-280-8832
Email: orders@isbs.com
Web: http://www.isbs.com

Asia and the Pacific (except Japan)
Kinokuniya Company Ltd.
Head office:
3-7-10 Shimomeguro
Meguro-ku
Tokyo 153-8504
Japan
Telephone: +81-(0)3-6910-0531
Fax: +81-(0)3-6420-1362
Email: bkimp@kinokuniya.co.jp
Web: www.kinokuniya.co.jp
Asia-Pacific office:
Kinokuniya Book Stores of Singapore Pte., Ltd.
391B Orchard Road #13-06/07/08
Ngee Ann City Tower B
Singapore 238874
Telephone: +65-6276-5558
Fax: +65-6276-5570
Email: SSO@kinokuniya.co.jp

ISBN 978-1-925608-99-1 (hardback)

ISBN 978-1-925608-14-4 (paperback)

Contents

Figures

Tables

Photographs

Contributors

KAWAI, Shuichi
Professor, Graduate School of Advanced Integrated Studies in Human Survivability, Kyoto University
Major field: Forestry, Wood Science, Environmental Science

FUJITA, Masakatsu
Professor, Graduate School of Advanced Integrated Studies in Human Survivability, Kyoto University
Major field: Philosophy, Japanese Philosophy

KAWAI, Ishida Eriko
Professor, Graduate School of Advanced Integrated Studies in Human Survivability, Kyoto University
Major field: Investment Management, International Management

HASHIGUCHI, Michiyo
Former Professor, Graduate School of Advanced Integrated Studies in Human Survivability, Kyoto University
Major field: International Cooperation

IALNAZOV, Dimiter Savov
Professor, Graduate School of Advanced Integrated Studies in Human Survivability, Kyoto University
Major field: Institutional Economics, Economic Development

IKEDA, Yuichi
Professor, Graduate School of Advanced Integrated Studies in Human Survivability, Kyoto University
Major field: Data Science, Network Science, Computational Science

IZUMI, Takura
Professor, Graduate School of Advanced Integrated Studies in Human

Survivability, Kyoto University
Major field: Archaeology, Ancient Civilization

KANAMURA, Takashi
Associate Professor, Graduate School of Advanced Integrated Studies
in Human Survivability, Kyoto University
Major field: Energy Finance, Risk Management

KIMURA, Senichi
Professor, Graduate School of Advanced Integrated Studies in Human
Survivability, Kyoto University
Major field: International Cooperation, Development Education

MITSUYAMA, Masao
Professor, Graduate School of Advanced Integrated Studies in Human
Survivability, Kyoto University
Major field: Microbiology, Infectious Diseases, Immunology

OHISHI, Makoto
Professor emeritus, Kyoto University
Major field: Constitutional Law and Legislation

OSHIMA, Koichiro
Professor, Graduate School of Advanced Integrated Studies in Human
Survivability, Kyoto University
Major field: Organic Chemistry, Organometallic Chemistry

SAKURAI, Shigeki
Professor, Graduate School of Advanced Integrated Studies in Human
Survivability, Kyoto University
Major field: Global Energy, Mineral Resources Issue

SOWAKI, Hiroshi
Professor, Faculty of Sociology, Kyoto Sangyo University
Major field: Educational Policy Research

YAMAGUCHI, Eiichi
Professor, Graduate School of Advanced Integrated Studies in Human
Survivability, Kyoto University
Major field: Physics, Innovation Theory

YAMASHIKI, Yosuke
Professor, Graduate School of Advanced Integrated Studies in Human Survivability, Kyoto University
Major field: Water Resource Engineering, Water Environmental Engineering, Earth & Planetary Science

ZHAO, Liang
Associate Professor, Graduate School of Advanced Integrated Studies in Human Survivability, Kyoto University
Major field: Fundamental Informatics, Computation

Integrated Studies in Human Survivability

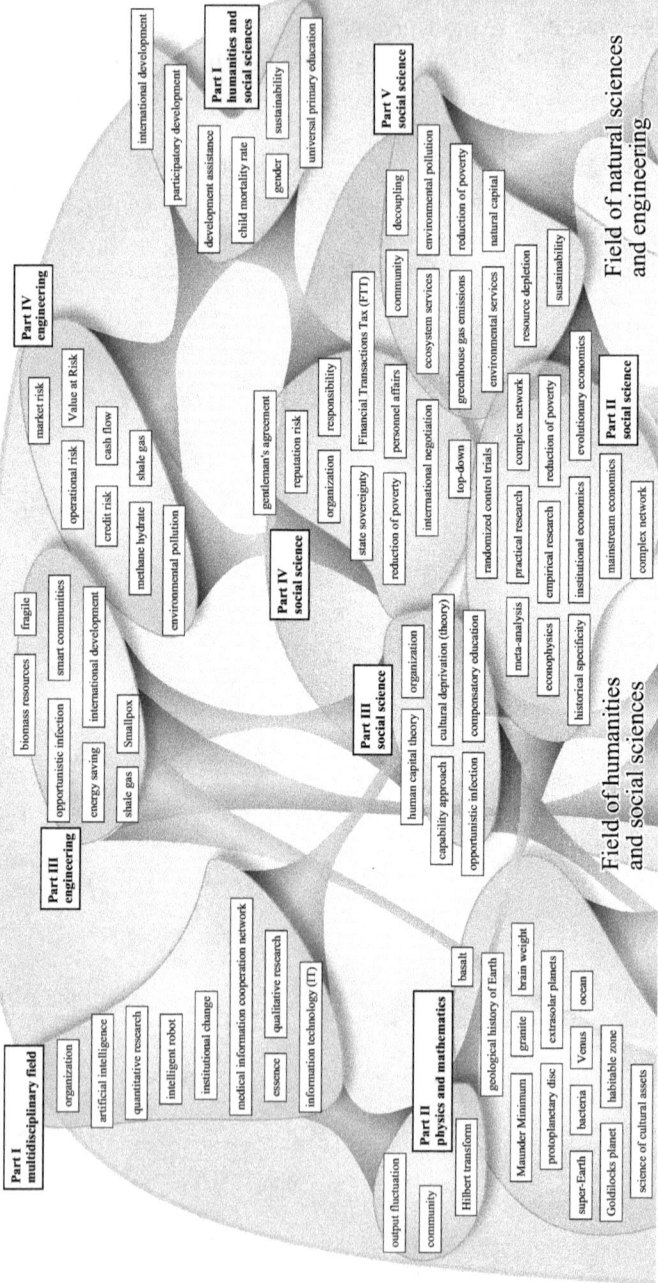

Part I multidisciplinary field

- output fluctuation
- community
- Hilbert transform
- organization
- artificial intelligence
- quantitative research
- intelligent robot
- institutional change
- medical information cooperation network
- essence
- qualitative research
- information technology (IT)

Part II physics and mathematics

- geological history of Earth
- basalt
- brain weight
- Maunder Minimum
- granite
- extrasolar planets
- protoplanetary disc
- Venus
- ocean
- super-Earth
- bacteria
- habitable zone
- Goldilocks planet
- science of cultural assets

Part III engineering

- biomass resources
- fragile
- opportunistic infection
- smart communities
- energy saving
- international development
- shale gas
- Smallpox

Part III social science

- human capital theory
- organization
- capability approach
- cultural deprivation (theory)
- opportunistic infection
- compensatory education

Part IV engineering

- market risk
- Value at Risk
- operational risk
- cash flow
- credit risk
- shale gas
- methane hydrate
- environmental pollution

Part IV social science

- gentleman's agreement
- reputation risk
- responsibility
- state sovereignty
- Financial Transactions Tax (FTT)
- organization
- reduction of poverty
- personnel affairs
- international negotiation
- top-down
- randomized control trials
- meta-analysis
- practical research
- complex network
- econophysics
- empirical research
- reduction of poverty
- institutional economics
- evolutionary economics
- historical specificity

Part II social science

- mainstream economics
- complex network

Part I humanities and social sciences

- international development
- participatory development
- development assistance
- child mortality rate
- sustainability
- gender
- universal primary education

Part V social science

- decoupling
- environmental pollution
- community
- reduction of poverty
- ecosystem services
- natural capital
- greenhouse gas emissions
- environmental services
- resource depletion
- sustainability

Field of natural sciences and engineering

Field of humanities and social sciences

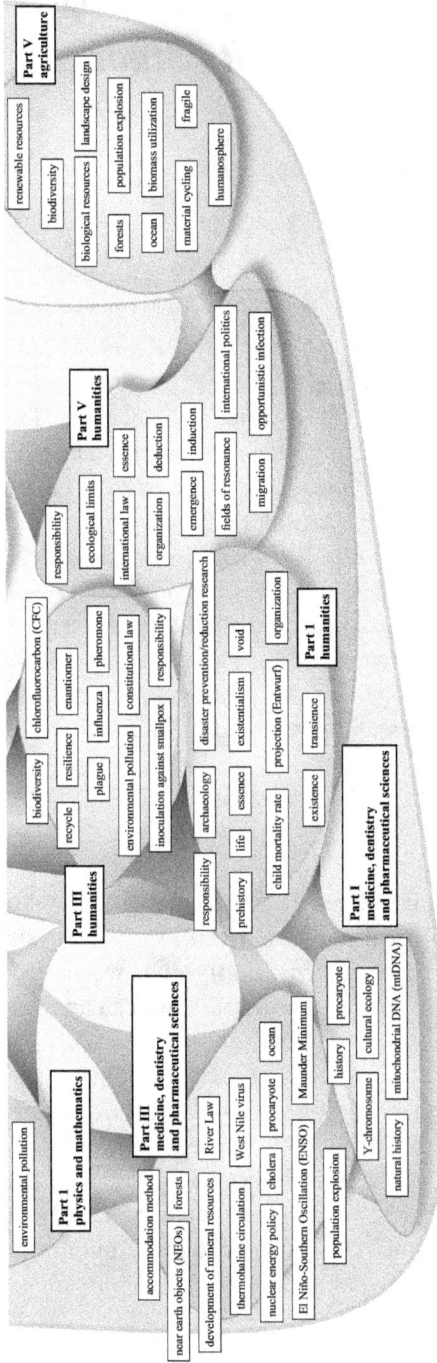

Part V agriculture

- renewable resources
- biodiversity
- biological resources
- landscape design
- forests
- population explosion
- ocean
- biomass utilization
- material cycling
- fragile
- humanosphere

Part V humanities

- responsibility
- ecological limits
- essence
- international law
- deduction
- organization
- induction
- emergence
- fields of resonance
- international politics
- migration
- opportunistic infection

Part III humanities

- biodiversity
- chlorofluorocarbon (CFC)
- recycle
- resilience
- enantiomer
- pheromone
- plague
- influenza
- constitutional law
- environmental pollution
- responsibility
- inoculation against smallpox
- responsibility
- archaeology
- disaster prevention/reduction research
- void
- organization
- prehistory
- life
- essence
- existentialism
- projection (Entwurf)
- child mortality rate
- transience
- existence

Part I humanities

Part I medicine, dentistry and pharmaceutical sciences

Part III medicine, dentistry and pharmaceutical sciences

Part I physics and mathematics

- environmental pollution
- accommodation method
- forests
- near earth objects (NEOs)
- development of mineral resources
- thermohaline circulation
- West Nile virus
- River Law
- nuclear energy policy
- cholera
- procaryote
- ocean
- El Niño-Southern Oscillation (ENSO)
- Maunder Minimum
- population explosion
- history
- procaryote
- Y-chromosome
- cultural ecology
- natural history
- mitochondrial DNA (mtDNA)

Explanation of the Network Analysis used in *Human Survivability Studies*

Introduction

In this book, we used the data science technique of text mining to process natural language, the results of which have been used to display the text as a network. On this basis, community analysis, a network science technique, has been conducted to extrapolate network characteristics. In particular, the relationships between the global issues discussed in *Human Survivability Studies*, those between the fields of study (sciences) that tackle these issues and those between global issues and science have been clarified. Finally, we presented the characteristics of Human Survivability Studies as 'collective intelligence', i.e. the amalgamation of Human Survivability Studies as interpreted by the authors of this book.

Morphological analysis

Here, the technique of text mining will first be explained, with particular focus on morphological analysis. In this form of analysis, text (unstructured data) is divided into individual words. By aggregating which words are used and how often, we can analyze the text statistically.

As an example, we subject the following text consisting of the two sentences to morphological analysis: 'In the midst of various sciences, Human Survivability Studies represents the latest collective intelligence. Collective intelligence has validity for the purpose of research on global issues'. These sentences have been broken down into the sequences of words shown in quotation marks below.

1st Sentence

'In (preposition)' + 'the (particle)' + 'midst (noun)' + 'of (preposition)' + 'various (adjective)' + 'sciences (noun)' + ', (punctuation)' + 'Human Survivability Studies (noun phrase)' + 'represents (verb)' + 'the (particle)' + 'latest (adjective)' + 'collective intelligence (noun phrase)' + '. (punctuation)'

2nd Sentence

'Collective intelligence (noun phrase)' + 'has (verb)' + 'validity (noun)' + 'for (preposition)' + 'the (particle)' + 'purpose (noun)' + 'of (preposition)' + 'research (noun)' + 'on (preposition)' + 'global issues (noun phrase)' + '. (punctuation)'

Here, the '+' symbol indicates that each word in a sequence belongs to the same sentence. Next, the nouns and noun phrases will be extracted from these sequences.

1st Sentence

'midst (noun)' + 'sciences (noun)' + 'Human Survivability Studies (noun phrase)' + 'collective intelligence (noun phrase)'

2nd Sentence

'collective intelligence (noun phrase)' + 'validity (noun)' + 'purpose (noun)' + 'research (noun)' + 'global issues (noun phrase)'

By treating these nouns as nodes and making links between nodes in the same sentence, these sequences of nouns and noun phrases can be expressed as a network.

'midst' – 'sciences'
'midst' – 'Human Survivability Studies'
'midst' – 'collective intelligence'
'sciences' – 'Human Survivability Studies'
'sciences' – 'collective intelligence'
'Human Survivability Studies' – 'collective intelligence'
'collective intelligence' – 'validity'
'collective intelligence' – 'purpose'

'collective intelligence' – 'research'
'collective intelligence' – 'global issues'
'validity' – 'purpose'
'validity' – 'research'
'validity' – 'global issues'
'purpose' – 'research'
'purpose' – 'global issues'
'research' – 'global issues'

Here, the symbol '–' signifies a link. The first six links indicate relationships in the first sentence, the remaining ten those in the second sentence. Here, it should be noted that the text comprising the two sentences is expressed as a single network. The 'collective intelligence' node is linked to the 'Human Survivability Studies' node belonging to the first sentence and to the 'validity', 'purpose', 'research' and 'global issues' nodes of the second sentence, and is therefore a node that bridges the two sentences.

Glossaries

Next, text mining using a glossary will be explained. In this analysis, the only attributes are 'global issues' and 'science.' Before the analysis, a glossary recording the attributes of nouns forming nodes will be created, as shown below.

Global issues
'global issues (field A)'

Science
'sciences (field a)'
'research (field b)'
'collective intelligence (field c)'
'Human Survivability Studies (field c)'

The three nouns – 'midst', 'validity' and 'purpose' – are not included in either glossary. Using the two glossaries, the following networks consisting of nodes given the attributes of 'global issues' or 'science' are obtained.

'sciences (Science: field a)' – 'Human Survivability Studies (Science:)'

'sciences (Science: field a)' – 'collective intelligence (Science: field c)'

'Human Survivability Studies (Science:)' – 'collective intelligence (Science: field c)'

'collective intelligence (Science: field c)' – 'research (Science: field b)'

'collective intelligence (Science: field c)' – 'global issues (Global issues: field A)'

'global issues (Science:)' – 'research (Science: field b)'

The glossaries for global issues and sciences used in the actual analysis were created by asking the authors of each chapter to name ten keywords for each attribute of 'global issues' and 'sciences' used in their chapter, and to designate their attributes (Global issues / Science: Fields) and then collect those attributes and keywords from the book as a whole. Here, the authors were asked to designate fields for 'global issues' and 'sciences' from options prepared in advance. At the same time, these keywords were also used as index keywords after removing the attributes (Global issues / Science: Fields). As an example, a glossary for global issues in Chapter Six is shown in Table A.1 and a glossary for sciences in Table A.2.

The same procedure was repeated for all sentences in the book to express the whole book as a network. To make the network map easier to read, the respective networks for global issues and sciences were shown separately, and attention was paid only to links between nodes with large numbers of links between the two networks.

Community analysis of networks

Finally, the network science technique of community analysis will be carried out and network characteristics extrapolated. Community analysis is a method of extracting parts that are closely linked in a network by maximizing the modularity function. Here, the modularity function is defined as the difference between the total number of links made between nodes belonging to the same community, and the total number of links between nodes belonging to the same community in

Table A.1: Glossary of global issues

Keywords	Field of global issues
Globalization	International relations
Business cycle	Economic growth
Economic growth	Economic growth
International trade	Economic disparity
Economic shock	Economic growth
Greenhouse gas	Climate change and natural disaster
Climate change	Climate change and natural disaster
Energy security	Energy security
International organization	International relations
Economic disparity	Economic disparity

Table A.2: Glossary of sciences

Keywords	Field of science
Econophysics	Multidisciplinary field
Data analysis	Informatics
Synchronization	Mathematical and physical sciences
Hilbert transform	Mathematical and physical sciences
Renewable energy	Environmental science
Output fluctuation	Environmental science
Principle of maximum entropy production	Mathematical and physical sciences
Meta-analysis	Informatics
Complex network	Social science
Community	Social science

which links have been made at random so as to preserve the order. To find the community division that maximizes the modularity function, we start first from a state in which each node corresponds to a single community. We remove two from the total of all communities and calculate the modularity function on the assumption of a single community. We calculate the modularity function for all combinations, and adopt the combination in which the value of the modularity function is largest. We repeat this procedure, and finish the calculation when the modularity function stops increasing. In this way, we obtain the community division that maximizes the modularity function. On the network maps, nodes inside the shaded fields surrounded by the closed curves belong to the same community.

What can be read from the various networks

1. Network map of Human Survivability Studies
Human Survivability Studies are collective knowledge structured such that the scientific disciplines surround the liberal arts disciplines positioned in the center.

2. Part I network map
Global issues: Economic growth, ethics for life and the environment, climate change and natural disaster
Science: Physics and mathematics, medicine, dentistry and pharmaceutical sciences, humanities
Relationship between global issues and science: Survival – essence, Snowball Earth – habitable zones, human beings – procaryotes, cooling – history

3. Part II network map
Global issues: Ethics for life and the environment, economic growth, energy security
Science: Social science, physics and mathematics, multidisciplinary fields
Relationship between global issues and science: Poverty – evolutionary economics, human beings – meta-analysis, poverty – information technology

4. Part III network map
Global issues: Ethics for life and environment, economic disparity, poverty
Science: Engineering, humanities, social science, medicine, dentistry and pharmaceutical sciences
Relationship between global issues and science: Human beings – oceans, United Nations – international studies, economic growth – human capital theory, humanity – cholera, population growth – environmental pollution, human beings – energy saving

5. Part IV network map

Global issues: International relations, climate change and natural disaster, ethics for life and the environment

Science: Engineering, social science, humanities and social sciences

Relationship between global issues and science: International cooperation – gender, globalization – development assistance, human beings – market risk

6. Part V network map

Global issues: Ethics for life and the environment, economic growth, forest destruction

Science: Social science, agriculture, humanities

Relationship between global issues and science: Food – oceans, international law – international law, environment – forests, living beings – induction

Yuichi Ikeda

Introduction: Human Survivability Studies

Shuichi Kawai and Koichiro Oshima

This Introduction outlines the new field of Human Survivability Studies (HSS) that aims to address and resolve the complex structural issues now facing the global community. The integrated studies we produce are transdisciplinary and cover global issues with a broad spectrum, involving large-scale and complex systems such as cultures, industries, economies and nations.

The Introduction also describes the educational philosophy of the Graduate School of Advanced Integrated Studies in Human Survivability (GSAIS) of Kyoto University. This program has been designed to cultivate the next generation of leaders through the practical application of HSS.

What is 'Human Survivability Studies'?

How does Human Survivability Studies differ from the established sciences? Why is this field necessary at this point in time? The aim of this book is to construct a new scientific paradigm. It will first outline the social and scientific backgrounds, then clarify the purpose of HSS and discuss its framework.

Throughout the history of our planet and process of our biological evolution, many species have become endangered or extinct. Human history and the history of civilization also tell us much about the rise and fall of states and societies and local environmental change. In the same way, we can safely predict that the human race and our global systems, though currently thriving more than ever, will also face various problems and crises in the near future – and some of them will threaten our very existence. The same is true for states and societies. Therefore, HSS is a science that incorporates geological and biological history, human history and civilizational history, and uses this knowledge to map out means of survival for humankind and global society. For human beings, the realities of life and death are both unfathomable and inescapable; they have posed some of our most important philosophical conundrums since time immemorial. But the survival of humanity and states or societies (organizations) poses another important challenge. To deal with this kind of challenge, it is necessary to pool knowledge from many different sources. Only by using a broad range of knowledge and understanding will we be able to solve problems, identify challenges and explore and enact methods of resolving them.

Contemporary society at various levels – from individuals to communities, states, global systems and global society – faces a compound set of problems. HSS is a practical science that tackles and attempts to solve these issues. This new academic approach presents a scientific system that integrates and structures the knowledge and wisdom of individual sciences. Over time, the various scientific disciplines have been developed and strengthened, but at the same time have also become compartmentalized. HSS rearranges and recomposes them, so that they can be used as knowledge for survival in practical situations.

This book discusses new ideas and scientific methods that will form the basis of human survivability. Further, in referring to practical applications, it attempts to build a framework for this new scientific system using case studies of initiatives on specific issues as well as exploration and innovation aimed at the future.

Challenges for contemporary society

The challenges we face today are growing conspicuously broad in scale and complex in nature. With the rapid growth of the world's industrial economies since the twentieth century, environmental problems have expanded from issues of local pollution to climate change, atmospheric pollution and other global-scale concerns. The 'Arab Spring' that started in Tunisia in 2010 before spreading to other North African countries has morphed into conflict and chaos in the Middle East. The US financial crisis triggered by the Lehman Brothers collapse of 2008 sparked a fiscal crisis in Europe and a global economic crisis. Japanese society, too, has rapidly lost its dynamism, due to the economic stagnation since the collapse of the bubble economy in 1992, the emergence of population shrinkage and the super-aging society, among other problems. In particular, Japan was dealt a heavy blow by a combination of natural and manmade disasters, namely the massive earthquake and tsunami known as the Great East Japan Earthquake of 2011 and the ensuing nuclear power plant accident. These disasters have led to a major loss of trust in science and technology, and the fragility of Japanese society was arguably exposed in the underdevelopment of its crisis management resources. On the other hand, the disasters also re-awoke awareness of the resilience of Japanese society in terms of the autonomy of social order and the strength of public spirit. They also underlined the national character of the Japanese, who act in observance of courtesy and order. Besides achieving an early recovery from the disaster, measures to thoroughly investigate the causes and use the results to mitigate future disasters could be framed as the most pressing issue.

Thus, the global community today faces a variety of challenges, including regional conflict, economic crisis, population problems, environmental pollution and the spread of infectious disease, and it is difficult to see how these issues can be overcome. The globalization and 'flattening' of information as well as people and goods have contributed greatly to the increasingly complex nature and broad geographical scale of problems. We are in the process of building an information infrastructure whereby anyone, anywhere in the world, can obtain homogeneous information simultaneously, albeit with differences in degree. By contrast, the trends toward local uniqueness, diversity and cultural tradition have emerged in reaction or opposition to the series of moves toward globalization. It is imperative that we achieve a paradigm shift with an eye on the future, while attempting to harmonize these changes.

Challenges for the sciences

How is the academic world planning to address these diverse challenges facing contemporary society?

Modern sciences have accumulated knowledge on humanity, society and nature, taking ancient Greek philosophy and sciences as their source. In particular, the modern sciences that have developed since the Renaissance in Europe have universalized science as a methodology, and have developed means of clarifying cause-and-effect relationships via reduction to simple elements and models. Scientific methods have arguably provided the foundations that support a wide variety of empirically based academic pursuits by accumulating facts through observation, experiments, investigation and other means of hypotheses testing. This method of exploring scientific wisdom by 'learning how to learn' has been extremely effective as a methodology for the continuous production of knowledge. As a result, it has been applied and adapted to many academic fields and has led to major achievements in various fields of natural science, including the elucidation of natural phenomena and the development of technology that enriches human life, in particular.

However, it has occasionally been pointed out that the sciences that have developed and intensified on the foundation of scientific methods have now deviated markedly from the actual situations and realities of human life and society, as well as natural phenomena. As a result, it could be argued that these foundations no longer adequately provide the means for a given scientific field to grasp phenomena in the complex and interdependent challenges facing contemporary society, or to elucidate and address their causes.

Regarding the Great East Japan Earthquake, mentioned above, issues have been identified in terms of the lack of organization and comprehensive analysis of previous knowledge and historical realities of earthquakes and tsunamis, as well as the inadequate use of measures to avoid risk and reduce damage. If we include physically imperceptible tremors, earthquakes occur tens of times per day in Japan and the surrounding region. Conversely, large earthquakes that cause significant ground motion and tsunamis due to tectonic movements, etc., occur only once every 100 to 1,000 years. In this way, the frequency and scale of earthquakes are known to follow a pattern of power distribution. How are we to confront the risk of a major earthquake that occurs only once every 1,000 years? How should we evaluate and manage the risk posed by earthquakes, determine standard criteria and devise disaster mitigation measures? It goes without saying that decision-making in this regard must be based on scientific knowledge. There is, however, no single solution to these challenges. It is therefore essential to involve residents of the areas in question in this decision-making process, as well as expert groups representing the government, industry and academic societies. The individual scientific fields engaged in addressing these challenges are wide-ranging; they include not only technical fields such as physics and engineering, but also law, economics, the social sciences, philosophy, history and other fields in the humanities.

The Great East Japan Earthquake also exposed a number of socioeconomic issues that need to be considered. Aside from technical issues related to preventing accidents in nuclear power plants, these also include directions for risk management and power supply systems, knock-on effects to resources and energy policy, the economy and industry and ways of achieving a process of consensus through resident participation. With a view to reconstructing local residents' lives following the Fukushima nuclear power plant accident, moreover, we urgently need to specify issues, identify aims and implement various measures based on medium- to long-term prospects. These include the technical problems and financial burdens involved in removing radiation from polluted areas, the technical and social issues related to the storage of radioactive waste and pollutants, the evaluation of risk and creation of management standards based on scientific evidence and the dissemination of scientific literacy. It is clear that we cannot adequately address such a complex disaster from individual scientific fields alone.

Many disciplines and fields have developed independently, and the interconnections and networks between fields have been lost in the process. As a result, the sciences seem to have become isolated in their fields, lost their flexibility and become detached from a society in a period of major transition. This has made it harder to address real social issues from narrow specific fields alone, as they have become segmented along with their increasing sophistication.

In order to resolve this, a flourishing trend in recent years has seen researchers from numerous fields become involved in interdisciplinary and multidisciplinary research. The growth of interdisciplinary research, which spans the boundaries between fields and thus expands their scope, will lead to the rise of new specific interdisciplinary fields, and these will eventually become recognized as independent disciplines. If we step back and look at the bigger picture, however, we may also perceive this trend as one that promotes segmentation. Thus, if the development of sciences is overly focused on establishing narrow specific fields, this process could swallow the new fields in the pitfall of segmentation. What the sciences need now is a new approach to explore transdisciplinary fields through the process of reorganization, integration and emergence. This will enable us to gain an understanding of science as a whole, while connecting specific fields freely whenever necessary to form networks and solve social problems.

What are Survivability Studies?

As mentioned above, 'life' (the meaning of living) and 'death' in combination present human beings with our most important philosophical proposition. Yet until now, the history of the human race has arguably focused only on the path of development based on the concept of 'historical progress'. Today, we also need to consider the problem of human extinction and avoiding this outcome, along with the progress and development of the human race, global systems and the global community. We must understand the diversity of species and the uniqueness of states, societies, cultures and values, and create a world in harmony with the contemporary trends of globalization and the universalization of technology and information.

An important aspect of Survivability Studies is that we draw on past case studies to acquire knowledge, technology and systems to perpetuate evolution and diversity and avoid extinction for all living organisms including humans. These issues have not been sufficiently investigated until now. On this basis, we prepare for the future. The work of 'relativizing' the human race, as it were, is an important task.

Survivability Studies is a scientific approach that searches for ways to extend the existence of human beings and the global community. The studies we produce aim to acquire the wisdom needed to avoid seemingly inevitable crises. HSS is the collective name given to research that broadly explores ideas, methods, policies, applied technologies and other practices and applications aimed at overcoming society's problems. As such, it is not merely a question of theory and methodology, but has an inherently transdisciplinary and trans-science nature, with a broad perspective including a practical application to problem solving, case studies and the development of new innovations. We need to value the interconnections and links (i.e. context) between ideas or concepts for human survivability and the contents applied to put these into practice (methodologies, specific scientific fields, case studies, etc.), and to integrate these from an overarching perspective.

The research carried out under the umbrella of HSS aims to produce new knowledge for solving problems, i.e. knowledge and understanding for survival. This is achieved by taking a broad

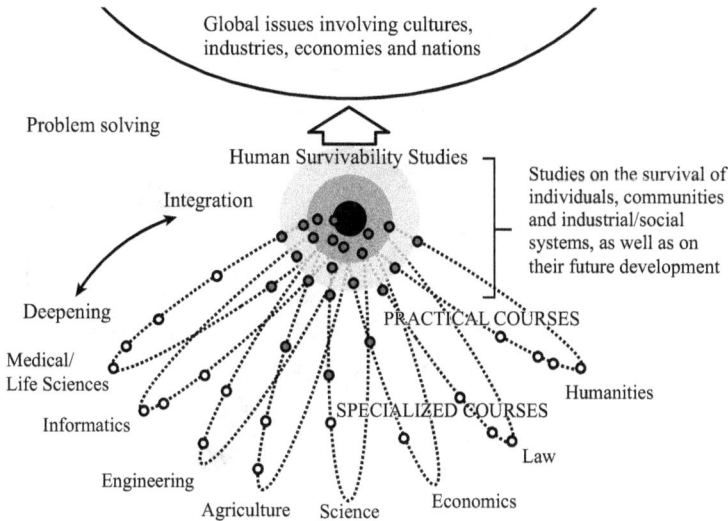

Figure 0.1: The relationship between Human Survivability Studies and existing fields of science

perspective of existing scientific systems, accumulating logical thought and bundling together, reorganizing, utilizing and merging specific fields. As such, HSS, as shown in Figure 0.1, may be perceived as a composite science that focuses on problems in relation to survival.

Essentially, there are two ways of looking at things – from a 'bird's eye view' or from a 'worm's eye view'. A bird's eye view is particularly relevant during times of change, or when it is difficult to predict what lies ahead. This view entails taking a macro-perspective, an overview of things. This is the 'eye' with which we form a composite picture of various events and phenomena, and combine them to identify future directions. The concept behind 'HSS' can be said to rely on this kind of perspective. But even if we can use a bird's eye view to forecast the signs of future shifts, we then need contingency plans with which to address them. That's why we also need a worm's eye view. This is the 'eye' used to ascertain diversity and individuality in more detail from a micro-perspective. It is not simply enough to have a 'grand design' when creating systems; fine-tuning with close attention to detail is also necessary,

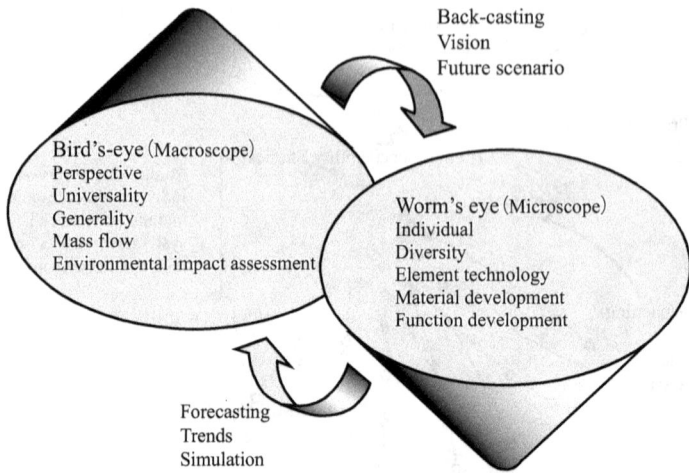

Figure 0.2: Bird's eye view and worm's eye view

and for this it is essential to have new technologies and systems. To put it succinctly, HSS broadly takes a bird's eye view to issues, while individual disciplines provide a worm's eye view. The two are in a mutually complementary relationship.

The issues tackled by Survivability Studies are social issues that have become complex and broad in scale, for example, population problems, environmental destruction, securing food, water and energy, addressing inequality, eradicating poverty, promoting access to education and tackling other immediate and pressing issues related to human survivability that are shared worldwide. Then there are other issues connected with our response to safety and security and health and crisis management, such as regional conflicts, terrorism, extreme weather, disasters and infectious diseases. One could also cite issues specific to certain regions; in Japan's case, for example, we have a super-aging society, the depopulation of mountain villages, regional revitalization focusing on small and medium-sized cities and measures to combat earthquakes, to name a few. Although these are mainly local problems, they could equally be viewed as issues that will one day impact on other parts of the world.

These social issues that confront us both domestically and globally suggest that the natural, ecological and social environments around us have become increasingly pluralistic, interdependent and complex as a consequence of globalization. Not only are these three environments mutually interrelated, but they are also prone to be impacted by violent change. This makes it difficult to address the problems facing the contemporary era sufficiently with existing knowledge and experience, or by forecasting based on an accumulation of existing situations alone. There are limits to methods of simulation based only on endorsing the status quo. Conversely, the use of goal-oriented 'back-casting' has now started to attract attention. Scientifically, back-casting remains little used at this point due to the difficulty in specifying parameters for setting the 'desirable future' (goal) to be targeted. It does however have clear targets as a practical method in real situations, and the effects of application are considerable. It is effective as a way of envisioning the desired future image of society, depicting scenarios aimed at materializing this, foreseeing changes, presenting a vision for the future and materializing goals. As such, it could be seen as a practical method.

The structure of HSS

Survivability Studies are conceptually very similar to 'Sustainability Studies'. The focus of the latter was on environmental problems, then expanded to the wide range of global issues being targeted as Sustainable Development Goals (SDGs) in 2030. The main concerns here include mitigating climate change in the geosphere and reducing carbon dioxide emissions, halting the depletion of resources and energy and conserving biodiversity in the biosphere and preserving food and water resources and achieving zero emissions in the human sphere. In 'Our Common Future' (1987), a report by the UN Brundtland Commission, 'sustainable development' is defined as 'development that meets the needs of the present without compromising the ability of future generations to meet their own needs'. In other words, it is defined as development that ensures intergenerational equality. In the UN Conference on Environment and Development (1992), 'environment and development' are set on an equal footing as two sides of the same coin. But in reality, environmental protection and development or growth are often in a trade-off relationship; the expression of sustainable development (growth) often places weight on development while environmental protection merely exists as a front.

Survivability Studies, as outlined above, is concerned with technology, policies, tactics and strategies designed to facilitate survival and the wide-ranging issues related to crisis response and decision-making. Both of these are conceptual in nature, and both take a macro-approach, such as attempting general analyses from a 'bird's eye view'. In HSS, however, existing frameworks (structures), functions and forms are expected not only to be maintained, but also to be freely transformed if necessary. As will be shown below, HSS is characterized by unified ideas, methodologies and practices for opening up future possibilities, including case studies on individual issues and innovation. This is a unique characteristic that differs significantly from Sustainability Studies and other fields of science.

Human Survivability Studies is a problem solving, goal-oriented scientific system that focuses on issues; it does not rigidly adhere to any specific discipline or field, but extracts the knowledge and experience needed to resolve issues from a broad spectrum, then explores and enacts measures to solve them. With this kind of dynamism, this new transdisciplinary field requires support from new ideas and philosophy. We need to form a new concept by combining words that express the ideas developed by Survivability Studies with heterogeneous words. At the same time, new methodological designs are also required, in line with the objective of this new field. We also need a theory of practice, including case studies, applied technology and policy research to relativize various sciences, gather and analyze the outcomes of a wide range of sciences and ultimately resolve issues.

Figure 0.3 shows the framework of ideas, methodologies and practice behind HSS in graphic form.

As is clear from the figure, the component elements of HSS are structured into three stages. Stage 1 represents the concepts, philosophical ideas, moral philosophy, historical studies and culture related to knowledge for survival. Stage 2 represents scientific methodology, analytical methods and logical thinking and Stage 3 involves practice, including policy, innovation, applied technology and case studies, as well as applied science. Taking Stage 1 as the concept and Stages 2 and 3 as the contents, sufficiently understanding the correlations and links connecting these two (context) is extremely important in Survivability Studies. This may be interpreted as a foundation for thinking about and practicing rational and appropriate methods of solving the problems that face

Human Survivability Studies includes developing the philosophy, methodology and applications to resolve complex social issues

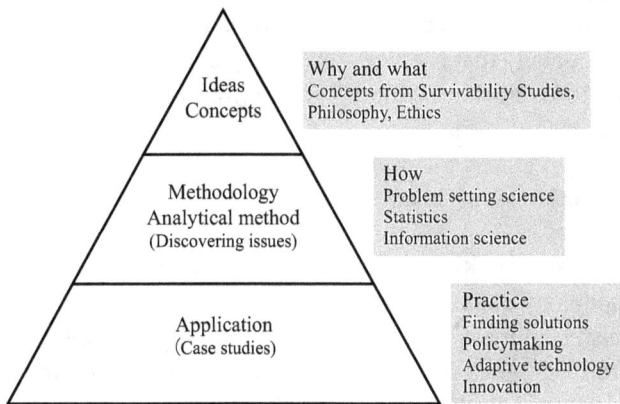

Figure 0.3: Framework of Human Survivability Studies

contemporary society – or in other words, as cultivation (intelligence, knowledge, practical ability).

The ideas applied to philosophy and science are characterized by aiming to solve problems continuously through a series of processes, namely discovering problems (establishing or extracting hypotheses), data mining by observing, investigating and experimenting, data analysis by creating models, etc., verifying and evaluating conclusions and reestablishing problems (hypotheses) (see Figure 0.4). Because scientific methodology can be applied to many issues, it has been established as a general research method in all kinds of academic domains across the natural sciences, humanities and social sciences. The PDCA (Plan-Do-Check-Action) cycle that is widely applied in industry and corporate strategy today can also be perceived as having brought scientific methods into the world of practical business, where they have developed into improvements and innovations in technology and organizational and corporate systems.

However, because scientific methods simplify and create models in their analytical process, they tend to fail to consider gaps in relation to reality. This sometimes causes them to lose sight of their original objective. In HSS, as shown in Figure 0.5, academic papers from various fields of science and documents such as reports by governments, public corporations, think-tanks and others are treated as metadata and case studies, and are aggregated and analyzed with the focus on a given problem. This means that science, statistics and information science form the methodological foundation. It is essential that, by integrating metadata and case studies, we develop a methodology that pursues universality and diversity and provides a structure of knowledge for survival, thereby providing a 'bird's eye view'.

Composition of this book

This book is an attempt to show the framework behind the ideas, methodology and practice of HSS. HSS has only just begun life as a transdisciplinary field of science. By surveying the survival and future of the human race and the global community, identifying problems and fleshing out comprehensive solutions, case studies, practical research and so forth, we hope to create a balanced constitution and structure in the near future.

This book consists of five parts. Here in the Introduction, we have raised the question 'What is Human Survivability Studies?' and discussed the objectives and framework behind this new field, thereby outlining the overall structure of this book.

In Part I, we discuss actual issues related to human survival founded in the humanities, including the ideas behind Survivability Studies emerging from philosophy, literature, historical studies, etc. We begin by questioning what the survival of human beings and the human race actually constitutes and consider the meaning and significance of survival by taking a bird's eye view of our planet, life and the paths of development, maturity and decline of the human race. We also examine language and communication as the foundation for cross-cultural and multi-functional understanding.

In Part II, we attempt to build a methodological basis for Survivability Studies, taking into account approaches and methods used in the social and natural sciences. We also discuss the advanced techniques used in information technology, a field that has seen such conspicuous growth in recent decades, and the approach used by Survivability Studies to apply these techniques.

Parts III to V are concerned with the application of Survivability Studies. They introduce case study research showing the bird's-eye view adopted by Survivability Studies regarding particular issues and its general approach to solving problems. Specifically, Parts III

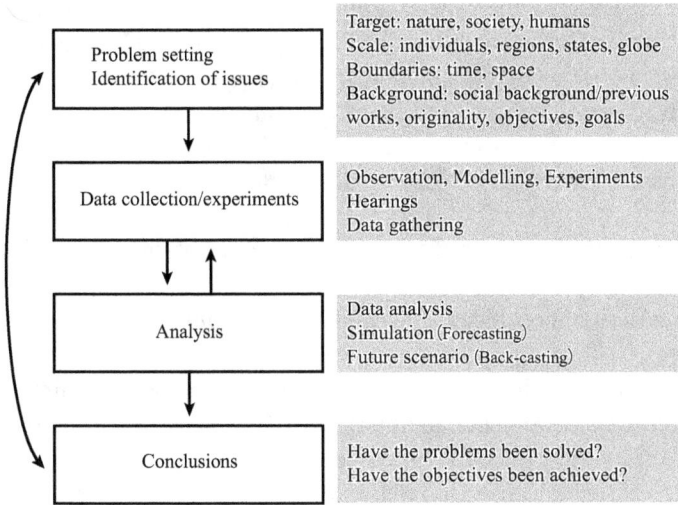

Figure 0.4: Methods used in philosophy and science: Ways of exploring knowledge, 'learning how to learn'

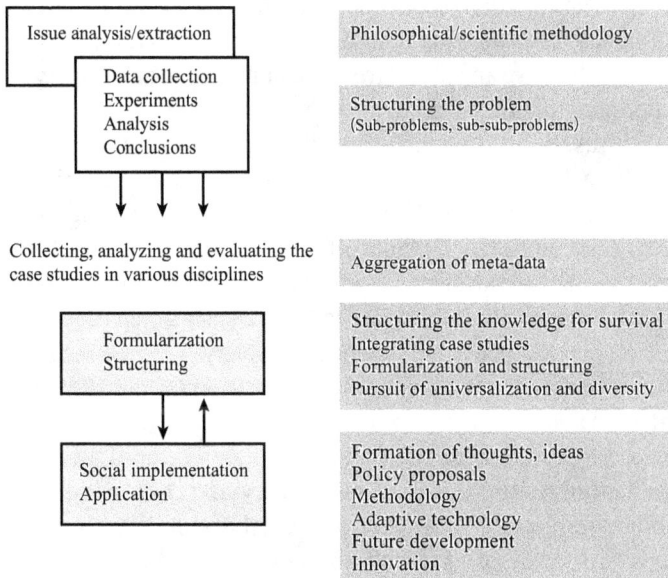

Figure 0.5: Methods used in Human Survivability Studies

and IV deal with the various problems facing contemporary society and issues for Survivability Studies as practical fields, while Part V raises issues related to opening up future possibilities.

We hope this book will be of assistance to readers in helping them understand the challenge of building the new academic transdisciplinary field of Survivability Studies.

Educational philosophy on Human Survivability Studies

Human resources today: Developing leadership material

While scholarship in individual fields of specialization is aimed at specific fields, objects, methods, etc., HSS specifies no field and targets a variety of tasks. To compare it to a jigsaw puzzle, individual fields of specialization in science could be seen as the pieces, while HSS reveals the whole picture, clarifies mutual meanings among the pieces and the relationships between them and creates a composite picture (overall solution) from the pieces. In HSS, we present overall prospects and visions for complex issues, and use and practice multiple individual fields of specialization to clarify methods of analyzing and solving issues.

Prior to now, the point has often been made that, while subdivision has furthered the development of sciences in isolation, it has also increased their estrangement from the human race and society; they are no longer equipped to deal with the contemporary issues that confront us. Schrödinger, for example, states that the many and varied branches of academic disciplines have expanded significantly both in breadth and depth, but that the only way of linking the sum of their parts into a single whole is to embark proactively on the work of integrating the various facts and theories (1944). In recent years, the social responsibility of science and scientists was brought into question and the need for 'Science in society, Science for society' was extolled under the Budapest Declaration (at the 1999 World Conference on Science).

As a way of coping with issues such as the rapid advance of globalization in the twenty-first century and the various global problems and crises that rarely occur but entail serious risks, activity in fields of research and education addressing global studies, leadership studies or existential risk has become animated in universities and research institutes all over the world. In Japan, too, the Great East Japan Earthquake disaster and ensuing nuclear power plant accident of March 2011 demonstrated the importance of having strong leaders when coping with crises. On a global level, similarly, there are concerns over the lack of leaders able to pose solutions to increasingly complex issues. The search is on for leaders able to see things from a macro-perspective and develop ways to break through the sense of insularity permeating the current era.

Thus, developing leaders is now a major challenge both in Japan and abroad. In policy terms, the Japanese government has published the 'New growth strategy' (2010), and, in the educational sphere, the Central Council for Education has reported on 'Graduate school education in a globalized society: To make graduate school graduates more active in diverse fields around the world' (2011). Meanwhile, the Japan Business Federation (2011), the Japan Association of Corporate Executives (2012) and others in the business arena have also made similar proposals concerning the development of global human resources. In fact, the question of how leaders should be developed is one that is being broadly discussed in society.

The creation of the Kyoto University Graduate School of Advanced Integrated Studies in Human Survivability (GSAIS) represents a contribution to dealing with the issues raised above, from academic and educational perspectives. The aim of education at GSAIS is to 'develop global leaders' and to nurture individuals who will acquire the knowledge to link integrated understanding with practice, and will in turn connect practice to academic research. To put it another way, while graduate schools in Japan have until now trained individuals who have immersed themselves deeply in individual academic fields, the educational aims of GSAIS are to equip students with a broad liberal education and macro-perspective, and to train them to become a new type of expert practitioner able to take up the challenge of resolving issues in the global community.

Imagining and developing new leaders
Throughout industry, government and society at large, the issue of developing innovators and leaders is subject to a great deal of expectation. But what is the specific form of leadership that is actually needed and sought?

The desired image of a leader may differ depending on social circumstances, organizations, attitudes, etc. The qualities required of a leader may also change. In an epoch like the present, characterized by constant change, or when a shift to a new paradigm is occurring, it is very important to have the foresight to look ahead. There is a great weight of expectation on leaders able to not only present a vision, but also to enact and provide the driving force behind it. For this, various characteristics are required, such as decisiveness, communicating from multiple points of view and guiding innovation, for example. In fact, if we look closely at the image of leadership sought by the world and society, it seems as if the focus is not necessarily fixed. This is probably because the characteristics required of a leader tend to be based more upon qualities or disposition than on knowledge. This makes it difficult to develop leaders using conventional methods of education, the aim of which is to teach knowledge and skills.

So, what should be done to develop this kind of leader? It is often said that 'Leaders are born on the battlefield'. This could be paraphrased as 'The important thing is not knowledge but practical experience'. In that sense, 'the battlefield' is a great teacher. It is by actually confronting issues in the field and having no option but to address them that those who take up the challenge of forging solutions will emerge. We could say that leaders are trained through actual practice. If there is one quality that is shared by leaders, it is a kind of strong will, a belief and determination to accomplish 'something', whatever that 'something' might be.

The Shishu-Kan challenge: Establishing GSAIS

In the spring of 2010, a decision was made to create a new type of graduate school at Kyoto University, following a proposal by Hiroshi Matsumoto, the University's President at that time. A taskforce was organized, and a concept for developing human resources together with an educational curriculum designed to materialize this concept were drawn up. Unlike conventional graduate schools, the educational goal was to 'develop next-generation leaders who will be active in the global community', or in other words, to develop individuals with a comprehensive, macro-perspective capable of becoming leaders in society. To achieve this goal, the Graduate School of Advanced Integrated Studies in Human Survivability was established three years later. Though the name is abbreviated to 'GSAIS', this Graduate School is commonly known as 'Shishu-Kan' because it takes care of the 'Kyoto University Graduate School

Shishu-Kan' human resource development program (a leading program in doctoral course education inaugurated in 2011–2013). In other words, the organization responsible for implementing the 'Shishu-Kan' is the Graduate School of Advanced Integrated Studies in Human Survivability.

Rather than being the name of an actual academic field, such as engineering or law, the 'Shishu' in 'Shishu-Kan' refers to the style and method of academic discipline targeted by GSAIS ('Kan' means a building or hall). The education is based on the Buddhist concept of *mon-shi-shū* (listening to teaching, thinking about its truth and mastering that truth through practice). This refers to the three stages of wisdom obtained through the pursuit of knowledge, namely *mon-e* (receiving wisdom), *shi-e* (thinking wisdom) and *shū-e* (practicing wisdom). Based on learning in university faculties (*mon-e*) and the knowledge and experience gained from it, GSAIS was named 'Shishu-Kan' as a platform for logically integrating this learning and engaging in deep thought to link it together (*shi-e*), attempting to apply the findings thus obtained to the various problems facing contemporary society, deepening and honing knowledge through practice (*shū-e*) and enacting this.

The GSAIS curriculum and its characteristics

Figure 0.6 shows a specific curriculum based on the concept of *shishū*. Being a graduate school, one of the main pillars is of course 'research'. Each student chooses a five-year research theme, studies subjects and conducts research to consolidate a foundation in a field of specialization, while also taking practical subjects such as leadership and doing an internship. At the same time, students also study a liberal arts subject called *Hasshi* (lectures on integrated academic foundations). The curriculum is designed to produce, over the five years of the course, individuals who can conduct research from multifaceted viewpoints using specialized knowledge; or in other words, practitioners who, at the same time as being experts, are also equipped with wide-ranging knowledge and the ability to apply it, and who are able to challenge social issues.

The GSAIS curriculum does not necessarily consist only of subjects with clear goals. For example, there is a subject called *Jukugi* (industry-government cooperation special seminar), the aim of which is to enhance abilities that cannot be directly taught. *Jukugi*

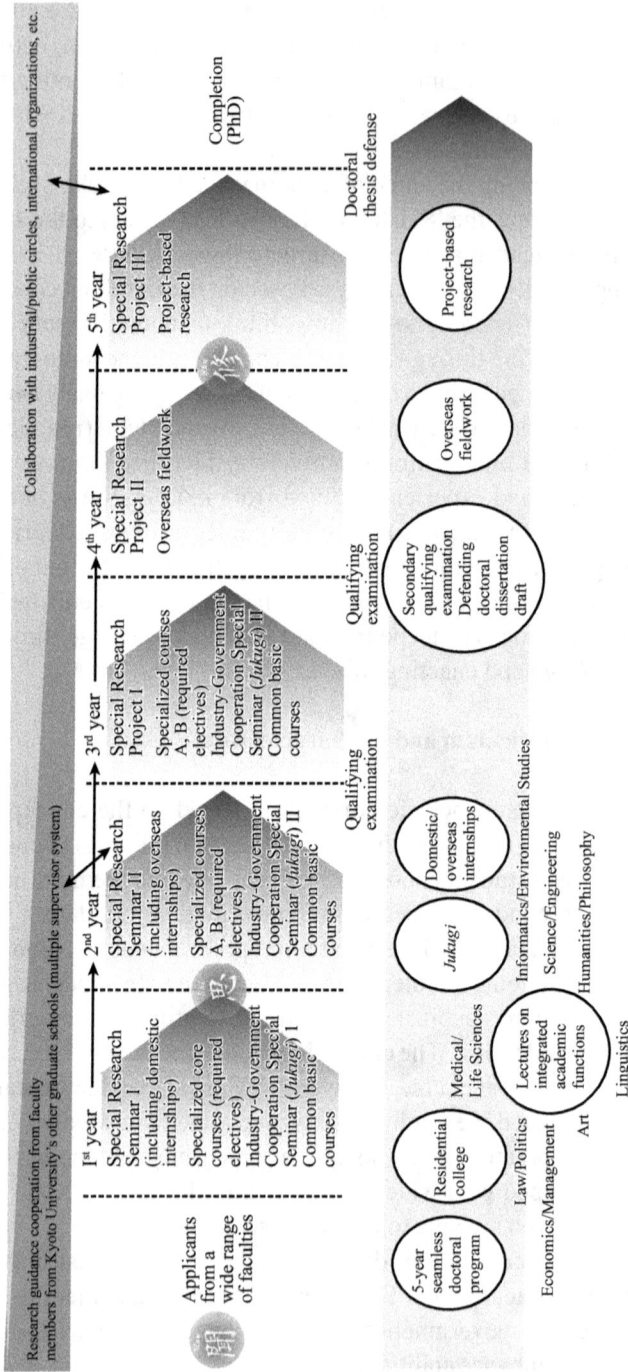

Figure 0.6: The curriculum at GSAIS

is an essential subject for training the leadership mindset, including self-awareness as a leader, determination, attitude, behavior and the spirit of challenge. Here, 'mindset' is particularly important. This is knowledge that comes from personal experience, as receiving tuition directly from someone with rich experience is the quickest way forward. This is a subject in which students are tutored directly by lecturers who are top leaders in various fields, including administration, industry, finance, international agencies, etc. They absorb their experience, feel their energy for themselves and make it their own. In this way, *Jukugi* gives students the opportunity to learn experience and wisdom from the field through dialog and Q&A with lecturers.

Internships and fieldwork to facilitate learning through practice are also incorporated into the curriculum. In the first year, students engage in domestic internships, or more specifically, volunteer activities in welfare institutions for the elderly. In the second year, they work as Short-term Overseas Youth Cooperation Volunteers for the Japan International Cooperation Agency (JICA) for one month during the summer recess. The details of the posting are determined after the submission of application documents, a written examination and an interview. Next, students study 'international cooperation', learn a local language and take part in preliminary training implemented by JICA. The volunteers sent to Bangladesh in 2015 carried out an organized and systematic impact assessment on a rural development project in cooperation with a local rural authority and the JICA office, with great success. This is a precious local experience for the students, as they experience homestays for one month in an environment where the language, customs, climate, culture and so many other things are different. In domestic and overseas internships, the basis lies in learning the spirit of service. Students acquire the ability to communicate, identify and solve problems, and take action through experience in the field.

Overseas fieldwork in the fourth year is designed for students to acquire international and social qualities through communication and exchange, to experience the processes of setting tasks and solving problems in the field and to hone their on-site and practical abilities. This is of course preparation for their future career path. Specific destinations include international agencies that will be their future places of employment, such as the Organization for Economic Co-operation and Development (OECD), the Food and

Agriculture Organization of the United Nations (FAO), the United
Nations Environment Programme (UNEP), the United Nations
Development Programme (UNDP) and the Economic Research
Institute for ASEAN and East Asia (ERIA), as well as research
institutes and global companies. Some students instead prefer to
work with international NGOs. Detailed plans are required in order
to connect with their future employment, while also linking to the
research content of their degree thesis. Although many students
currently want to work for international agencies, we expect more
students who will represent companies and organizations to emerge
in future.

In the fifth year, the students work on project-based research (PBR).
This is the culmination of their practical education and research up
to the fourth year. Students carry out PBR in organizations such as
companies, government bodies or local authorities. They plan and
implement projects in collaboration with their host organization, after
which they assess and verify the outcome before compiling a final
report. For example, they may hold international workshops or draw
up standards on environmental regulations in connection with the
policies of local authorities (although there is one student who aims
to start a micro-hydro power company during the fifth year). These
research projects provide opportunities to experience the sequence
of processes connected with a project and to hone the qualities and
abilities required of leaders from a comprehensive point of view.

To summarize the above, the Shishu-Kan Graduate School has
the following three characteristics. It introduces a tailor-made
curriculum in readiness for a five-year integrated degree program.
Enrolled students come from a wide variety of backgrounds,
including those from broad-ranging humanities, social and natural
science fields, those with previous professional experience and
international students. The research themes, subjects and fields
of specialization they tackle are also diverse. Their aspirations
and career expectations at the end of the five years are extremely
wide-ranging. This means that we form curricula in line with
each student's aspirations and career path, and in line with their
research themes.

For this purpose, we employ a teaching staff representing a wide
variety of fields, and adopt a system of multiple instructors based
on mentors and research instructors. These guide the students in
everything from care for their daily lives to deciding their study

subjects and monitoring their progress, while also giving detailed guidance on their research, etc. To achieve this, we operate a system of education in small groups. We also provide a residential college facility as a 'site' where students can share their daily lives, work hard and help each other while engaging in their studies and in deep thought. Finally, daily office hours have been set for the teaching staff at the residential college facility. The Graduate School of Advanced Integrated Studies in Human Survivability is a place that acts as an incubator, where young people from varied backgrounds and with diverse aspirations for the future can gather, take part in dialog and discussions also involving the teaching staff, foster mutual relationships of trust as friends and help each other to train.

The GSAIS theory of leadership

In the above, we have explained that the Graduate School of Advanced Integrated Studies in Human Survivability (GSAIS) has set the educational goal of developing leaders, with details of that educational goal, the curriculum and the educational environment. Here, we would like to introduce the GSAIS theory of leadership as summarized in Table 0.1. Attempts to develop specialists, practitioners and leaders who, though based in knowledge of specialized fields, can analyze and evaluate problems comprehensively by breaking through fragmentation or specialization, and who can make and implement practical proposals transcending individual fields to solve those problems, have started to appear mainly in western countries. 'Leadership' is also gathering attention academically. Until now, 'leadership' has mainly been a concern of companies training their employees through OJT (on the job training). In recent years, however, management and leadership have been discussed in business schools and other graduate institutes for specialized occupations, while efforts to develop global human resources and theories on leadership have also been thriving in domestic and overseas universities. A systematic pedagogy has been set in motion to this end.

At GSAIS, too, the necessary characteristics and mindset have been implicitly incorporated in subjects such as *Jukugi*, and a curriculum has been developed around these attributes. Skills with a direct bearing on leadership have been incorporated in overseas personnel training, internships and other practical curricula. The three lessons learnt in HSS – leadership mindset, leadership skills

Table 0.1: GSAIS theory of leadership (relationship between component elements and subjects)

---

1. Fostering the leadership mindset (human power)
→ (*Jukugi*)
   - Sense of mission, belief, spirit of challenge
   - Ability to set out a vision and goals: Ability to take a macro-perspective, foresight
   - Mindset: Ability to think flexibly, logically, critically

2. Acquisition of leadership skills (practical ability, ability to break through the field)
→ Domestic and overseas internships, overseas personnel training, PBR, foreign language education, management theory
   - Social attributes: Communication ability
   - International attributes: Understanding other cultures and diversity, language ability
   - Ability to motivate, ability to involve others, ability to mobilize others (teamwork, facilitation, coaching)
   - Management ability

3. Mastery of Human Survivability Studies
→ Special research seminars/special research projects, specialized core/specialized subjects, *Hasshi*
   - Solid specialized attributes
   - Wide-ranging knowledge of liberal arts

---

and the system of academic disciplines connected with these – are the core of GSAIS leadership theory. Leadership mindset means the mental attitude, way of thinking and spirit that a leader needs to have. It is a method of cultivating a sense of mission, belief and the spirit of challenge, training the ability to take a macro-perspective and the foresight needed to map out a vision and set goals and honing a flexible and logical way of thinking; it is a 'measure for developing human appeal'. On the other hand, leadership skills could also be called 'transferable skills'. In order to hone leadership skills, we need to deepen students' knowledge of other people, other cultures and diversity, and acquire international and social qualities, penetrative ability and practical ability. Communication and management are skills needed to lead a team in the field, motivate people and collaborate. With this mindset and these skills, leaders understand things more deeply, think subjectively, understand the viewpoints of other parties (i.e. stakeholders) and convey their own ideas in ways that are easy for others to understand. At the same time, they strive for change and creativity together with associates, and generate new value. An 'indomitable' spirit is also necessary. Enabling students to acquire a foundation for this ability and confidence is the aim of HSS. As we have already stated above, teaching this leadership mindset and these leadership skills means equipping students with the rationale and methods for overcoming problems, and the basic research ability and applied ability to explore the practice and application of policies, applied technology and other elements. This is a group of subjects designed to verify the students' own specialized nature and equip them with the ability to respond to the unknown by acquiring broad-ranging knowledge in liberal arts.

This book answers the question of what HSS should do as an academic approach for developing leaders. While this book is only at the stage of a guideline presenting a framework for this effort, we would like to flesh out this framework in future. We see it as an urgent task of higher education in Japan and the world to produce an innovative methodology for solving social issues, develop individuals who can lead practical initiatives and create an environment to facilitate this.

Part I
The Foundations of Human
Survivability Studies

economic growth

death

survival

food supply

gene analysis

energy

globalization

giant-impact hypothesis

critical solar flux

standard of living

environment

greenhouse gas effect

climate change

living being

disease

evolution

living organism

Snowball Earth

global issues

bacteria

infectious diseases

natural environment

human beings

oceanic anoxic events (OAE)

environmental adaptation

deforestation

cooling of weather global cooling

global warning

volcanic eruption

virus

air

threatened species endangered species

climate change and natural disaster

ethics for life and the environment

population growth

geological history of Earth

basalt

Maunder Minimum

protoplanetary disc

granite

brain weight

extrasolar planets

Venus

ocean

physics and mathematics

super-Earth

procaryote

bacteria

Y-chromosome

Goldilocks planet

habitable zone

history

medicine, dentistry and pharmaceutical sciences

science of cultural assets

natural history

cultural ecology

mitochondrial DNA (mtDNA)

environmental pollution

archaeology

responsibility

essence

organization

prehistory

existentialism

disaster prevention/ reduction research

science

life

child mortality rate

void

existence

transience

projection (Entwurf)

humanities

Introduction to Part I

Human Survivability Studies is a field of inquiry composed of various elements. These include conceptual elements related to the survival of individuals, society, nature and the environment, a vision for the future of society and the natural environment, methodological elements and elements related to specific problems and solutions. Human Survivability Studies emerges as a field of study when these elements are brought together organically. Part I of this text addresses the conceptual elements that provide the foundations for studies in this field.

In Chapter One, along with examining the essential question of what human beings are, we attempt to describe the direction we ought to be headed in, or, in other words, what the future of society and humanity *ought* to look like. In Chapter Two we address the question of how we can continue to live on this planet, or how we *ought* to go on living on this planet, from the perspective of the history of the Earth. In Chapter Three we retrace the history of life and make it clear that human survivability and preserving the Earth's environment are deeply intertwined with the preservation of all living organisms. In Chapter Four we clarify how human history has not been confined to human beings alone but has instead developed (declined) in close connection with the natural environment, and consider the meaning of this form of survival.

1 Questioning the Basic Nature of Human Beings: 'Where Have We Come from? What Are We? Where Are We Going?'

Masakatsu Fujita

Modern society is facing a broad array of problems, including the destruction of the environment, issues related to natural resources and energy, problems concerning population and food supply and wars and tension between states or ethnic groups. Human Survivability Studies (HSS) is a field of inquiry that addresses these difficult problems in which many factors are entangled, and, looking at them from a comprehensive, macro-perspective, brings together various disciplines in pursuit of solutions.

This field of inquiry is called *seizongaku* (survivability studies) in Japanese and Human Survivability Studies in English, but it is not the survival of individual human beings that is being examined; what is being addressed here is the survival of society, the survival of humanity as a whole and the survival of the environment, or the planet, itself.

Further, it is not simply 'survival' or 'continuing to exist' that is at issue; this field of inquiry will remain devoid of a core philosophy or set of ideas if we do not clarify the forms in which this notion is to be pursued. To put it another way, HSS forms a single discipline by forecasting the direction we are headed in, or, in other words, by taking on the task of describing how society, humanity and the planet ought to be in the future. This field obtains its foundations as a discipline by both seeking solutions to difficult problems from a technical perspective and also mapping out a vision for the future. This book constitutes an attempt to meet this challenge, and in this chapter I begin by demonstrating that the question 'Where are we headed?' is a fundamental topic of inquiry for all human beings.

Figure 1.1: Paul Gauguin, D'où venons-nous? Que sommes-nous?
Où allons-nous? *(1897–1898)*

When I think about 'where we are headed' in relation to survival,
what immediately comes to mind is a work the French artist Paul
Gauguin painted in Tahiti called *D'où venons-nous? Que sommes-
nous? Où allons-nous?* (Where have we come from? What are we?
Where are we going?).

It was painted during his stay in Tahiti from 1897 to 1898, but
it appears to be a depiction of Gauguin's internal world. Starting
from the viewer's right-hand side, first there is subject matter
symbolizing the beginning of life (a depiction of an infant),
next there is an image that symbolizes youth and the prime
of life and finally an elderly person on the verge of death is
portrayed. These correspond to the three questions in the painting's
title: Where have we come from? What are we? Where are
we going?

Of course, Gauguin was not the first to pose these questions –
they have presumably been asked since humanity took its first steps
towards self-awareness. The following passage is found in the gospel
of John (8:14): 'Jesus answered, "Even if I do bear witness about myself,
my testimony is true, for I know where I came from and where I am
going, but you do not know where I come from or where I am going"'.

Kamo no Chōmei's (1155–1216) well known *Hōjōki* (An account
of my hut) (1212) raises the same questions in its opening passage:

A river's flow is ceaseless, and yet its water is never the same. The
foam that froths in its eddies disappears and bubbles up, and does not
remain fixed there forever. ...Dying in the morning and being born

in the evening, [we are] just like the foam on the river. [We] do not
know, people who are born and die, where they have come from and
where they are going. (Kamo no Chōmei 1957)

Here the answer presented to the question of where people have
come from and where they are going is simply 'we do not know'.
This perspective makes the sense of transience that runs throughout
the entire work even more profound.

In philosophy, too, these questions have been taken up again
and again since ancient times as essential topics of inquiry. Pascal
(1623–1662) poses them in his *Pensées*. In a lecture included in
L'Énergie spirituelle (1919), Henri Bergson (1859–1941), one of the
leading French philosophers of the twentieth century, asks precisely
the same questions as Gauguin: 'Where have we come from? What
are we? Where are we going?' (Bergson 1959: 815) (Bergson no
doubt had Gauguin's painting in mind). Bergson asserts that while
systematic philosophy does not always directly engage with these
questions, they are what 'is perplexing, disquieting, and fascinating
for most men'. He goes so far as to say that if philosophy cannot
answer these questions, it is not worth an hour's effort.

'What are we?' Life and death

In the end, the questions illustrated by Gauguin can each be said to
be asking, 'What is a human being?' They are also asking about the
meaning of 'life'. As Bergson says, however, these are perplexing,
disquieting questions, and it is not at all easy to come up with clear
answers. The fact that they have been asked so relentlessly over the
centuries is another indication of this.

I cannot fully answer the question 'What is a human being?' but
there are two points I would make here. One is that we find ourselves
in a situation into which we have been placed through no intention
of our own, having been 'thrown' here as something that exists
in a particular state of being, and we have no choice but to accept
and deal with this fact. Martin Heidegger (1889–1976), a leading
German philosopher, described this state of affairs with the term
Geworfenheit (thrownness) (Heidegger 1972: 135). At the same time,
however, we are also entities that pursue our own potential within
this situation and the limitations and conditions it imposes on us. We
transcend our present selves in an ongoing process of self-creation.

This *Entwurf* (projection), to use Heidegger's term, is something that inherently belongs to human beings.

I would now like to consider the human trait of always being placed within a fixed set of circumstances. This amounts to simply thinking about what 'life' is, but with that which imposes a limit on 'life' as a clue to how we should proceed. In other words, the meaning of 'life' is brought into relief in the light of death as that which limits human existence.

Of course, we cannot talk about what 'death' itself is. In the *Apology*, Socrates says that death is something we do not know much about and cannot meaningfully discuss. Socrates had been put on trial and sentenced to death for seducing and corrupting the youth of Athens in his daily discussions with them ('rejecting the gods of the *polis* and believing in a species of *daimon*' was also given as a reason for his sentence). When Socrates was offered a chance to be freed on the condition that he would cease his discussions with the youth, an activity that to him was entirely the search for truth, or, in other words, 'to love and pursue wisdom' (φιλοσοφεῖν (*philosophein*)), he refused, saying that if he accepted such terms out of fear of his own death it would amount to a rejection of his own way of living – 'to love and pursue wisdom' whatever the consequences. He then told the people gathered for his trial that there was no need to 'fear death' because we do not know the first thing about it.

We are indeed ignorant about what 'death' is. We do know, however, that our lives are limited by it. We know that we are surrounded by a darkness, the nature of which we cannot grasp directly, and that the business of being alive is carried out within this framework.

Miki Kiyoshi (1897–1945) uses the term 'void' to express the situation of human life in such circumstances, that is, finitude or the fundamental condition of being unable to avoid death. In 'Ningen no jyōken ni tsuite' (On the human condition), an essay from his *Jinseiron nōto* (Notes on a philosophy of life), he writes, 'The more I try to focus myself, the more I feel I am floating above something. Above what? It can only be the void. My self is a point in the void' (1966: 254). A human being is like a tiny boat floating on the limitless sea of 'the void'. Miki believed it is this 'void' surrounding human beings that is the human condition, and if we ignore our relationship to it we will never be able to understand what we are.

Transience

Death, for us, is a darkness the nature of which we cannot grasp directly. This abyss whose bottom we cannot see causes us great anxiety. Since ancient times, people have faced the anxiety created by the ephemerality of existence and sought to express it. This sense of the transience of life can perhaps be described as the central theme that runs through the art, literature and religion of Japan.

One lucid interpretation of how people have faced death and transience is given by Karaki Junzō (1904–1980). Karaki is known as a literary critic, but he studied under Nishida Kitarō (1870–1945) and also wrote books on philosophy, such as *Miki Kiyoshi* (1947). In *Mujō* (Transience), a book published in 1965, after discussing the state of mind or emotion of 'ephemerality' displayed in classical Japanese women's literature, such as *Kagerō Nikki* (The mayfly diary) (ca. 975), *Genji Monogatari* (The tale of Genji) (ca. 1008) and *Izumi Shikibu Nikki* (The diary of Izumi Shikibu) (ca. 1008), and the 'pathos of transience' and 'awe-inspiring sense of transience' reflected in masculine emotions that appear in the works of male writers like Hōnen (1133–1212), Shinran (1173–1263), Yoshida Kenkō (ca. 1283–1352) and Bashō (1644–1694), Karaki focuses in particular on the 'metaphysics of transience' found in Dōgen (1200–1253).

Karaki thus distinguishes two kinds of 'transience' in this text. One is a 'sense of transience' that has been grasped as objects of 'mind', 'emotion' or 'awe', while the other is 'transience itself, a "metaphysics of transience" that gets right to the reality of things' (Karaki 1964: 352). This illustrates how death and transience have been addressed in two ways within the history of Japanese thought and literature. One approach has been to mourn the ephemerality of existence, observe one's own mind in the midst of dealing with this sorrow and carefully set down these observations in words, while the other has been to seek a way to live that cuts through the ephemerality or emptiness of existence without becoming drunk on one's own emotions. Karaki finds the latter approach in the thought of Dōgen.

The chapter entitled 'Shōji' (Life and death) in Dōgen's *Shōbōgenzō* (Treasury of the true dharma eye), for example, contains the following passage:

This life-death is the life of the Buddha. To loathe it or throw it away is to lose the life of the Buddha. To remain attached to life-death, too, is to lose the life of the Buddha, to stop the way of being of the Buddha in its tracks. Only when you neither hate nor love do you enter the mind of the Buddha. ...when you release and forget your own body and mind, throw them into the home of the Buddha, and follow what is done from the direction of the Buddha without applying your own force or using your own mind, you separate from life-death and become the Buddha. Should any person become stuck in his or her own mind? (Dōgen 1993: 468)

Dōgen is telling us to discard our grasping minds and avoid becoming attached to life-death. Here we are being told to walk down an entirely different path from a state of being in which we become intoxicated by a self that has been filled with feelings of 'ephemerality'. It is not that Dōgen never speaks about 'transience'. In another chapter of *Shōbōgenzō* (Treasury of the true dharma eye) entitled 'Dōshin' (Mind of the way), for example, he says, 'Turning our mind to transience, surely we should not forget the ephemerality of the world, and the precariousness of human lives' (Dōgen 1993: 471). But we are not being told this in order to lament this transience. On the contrary, it is precisely because the world is ephemeral that we are told to discard our attachment-prone mind and immediately separate ourselves from life-death. To do so, Dōgen believed, was to become a Buddha.

Returning to the question 'What are we?' or 'What is "life"?' the answer must surely be deeply connected to how we face the 'death' that imposes a limit on our lives. There is no single answer; various approaches can be taken. What can be said is that in contrast to Kamo no Chōmei's simple 'we don't know' in response to the question 'Where have people who are born and die come from, and where do they go?', Dōgen offers a clear answer to the question of where the tiny boat floating above the endless sea of 'the void' has come from and where it is going.

Existence

I have made two points above in relation to the question 'What are we?'. I've stated that on the one hand we have been thrown into a particular situation through no intention of our own, and have no choice but to accept and deal with it. At the same time, however, we are an entity that pursues its own potential within this situation; we are an entity that chooses its own state of being, and, going beyond its current state of being, is constantly creating itself. In what follows I consider this latter aspect of our being.

Here I draw on the philosophy of Jean-Paul Sartre (1905–1980), a philosopher whose existentialist ideas exerted a powerful influence on post-war thought. In *Existentialism is a Humanism*, Sartre distinguishes between the existence of things and the existence of human beings. Things are simply there, their existence nothing more than just being, while human beings, always aware of themselves, choose their own futures, make decisions and create themselves. In other words, human beings not only *are* in the manner of things, but *'exist'*. We throw ourselves into the future. Sartre borrows Heidegger's term *Entwurf* (projection) to describe this state of being of human beings. This *Entwurf* is our starting point, and there is nothing before it. Sartre expresses this as follows.

> Thus, there is no human nature since there is no God to conceive of it. Man is not only that which he conceives himself to be, but that which he wills himself to be, and since he conceives of himself only after he exists, just as he wills himself to be after being thrown into existence, man is nothing other than what he makes of himself. This is the first principle of existentialism. (Sartre 2007: 22).

With the phrase 'existence precedes essence', Sartre also expresses this idea that human beings, rather than possessing a fixed essence in advance and being determined by it, are entities that create themselves through their own decisions. This means that human beings are radically 'free'. The standards that justify our actions do not exist in advance outside of ourselves. We justify ourselves. But this means we take responsibility for everything. Our actions affect not only ourselves but also the people around us as well as society in its entirety. My choices 'bind' society and humanity as a whole.

As a result, I am responsible not only for myself but for other people and the entire human race. Because the choices I make are my own choices and not orders given by someone else, I must take full responsibility for them. And since everything starts with my own choices and decisions, I cannot escape this freedom and responsibility. Sartre expresses this as humanity having been 'condemned to freedom'. To run from this burden, and the anxiety it causes me, is to deceive, abandon and deny myself. We must accept the sentence we have been given.

Consideration for future generations

Returning to the question 'What is a human being?', a human being is an entity that freely decides and creates its future. But the choices human beings make are always tied to responsibility. In what form are we to bear this responsibility? This question is fundamentally connected to the 'where' of 'where are we going?'. Dōgen's answer describes a 'where' for oneself, but if one's actions are connected to every other person and 'bind' their manner of being, then the relationship between this 'where' and other people must also be examined.

In the past, ethics have taken as their object only other people who are right in front of us, or other people who are living in the same era. Even in cases in which people who will live in the future have been considered, ethical questions have been debated on the assumption that what is good for people in the present must be good for people in the future as well.

It cannot be denied, however, that issues related to ethics have undergone massive changes in the modern era. Our scientific and technological development has radically altered the circumstances in which we find ourselves; our activities no longer influence only the people around us, but, as can be seen in the case of global warming caused by the emission of greenhouse gasses, can affect the Earth as a whole. As Rachel Carson (1907–1964) warned in her book *Silent Spring* (1962), we are destroying the Earth's environment in an irreversible manner, and this is something that affects not only the present but also the distant future.

In the modern era, it has become impossible for us to discuss ethical issues without paying attention to future generations and the Earth's environment as a whole. Someone who has thought deeply about ethical issues from a new perspective in the midst of these circumstances is Hans Jonas (1903–1993), a philosopher who was born in Germany but spent his teaching career in Canada and the United States.

As its title suggests, in *Das Prinzip Verantwortung, Versuch einer Ethik für die technologische Zivilisation* (The imperative of responsibility: In search of an ethics for the technological age) (1979) Jonas conceives of a new ethics based on a 'principle of responsibility'. In previous conceptions of ethics people have been seen as having to make their own decisions and act in accordance with universal 'maxims' (principles established as guidelines for the actions of individual human beings). A classic example of this can be seen in Kant's fundamental ethical principle, which he referred to as a 'categorical imperative' in the sense that it was a rationally absolute imperative that was to be obeyed unconditionally.

Jonas makes significant changes to this principle, reformulating it as follows: 'Act so that the effects of your action are compatible with the permanence of genuine human life' (Jonas 1984: 11). Here what is being addressed is not what sort of principle or motivation you are acting in accordance with, but rather the effects your actions bring about. His 'categorical imperative' is that our actions must not only be such that they do not endanger the lives of the people of the future or their ability to live in a manner fit for human beings, but must moreover be such that they actively protect or guarantee them.

Since the continued existence of the Earth's environment and the survival of future generations are deeply connected to our actions, and since we have the potential to unilaterally determine their fate, we must also give careful consideration to the state of the environment and the survival and rights of future generations.

As was mentioned in the Introduction, the idea of 'sustainable development', that is, 'development that meets the needs of the present without compromising the ability of future generations to meet their own needs' has been presented in documents such as 'Our Common Future', the report issued by the United Nation's Brundtland Commission in 1987. In practice, however, more emphasis has been placed on 'development' than 'sustainability', and this continues to be the case today.

In these circumstances, taking into consideration the state of the Earth's environment and the survival and rights of future generations is our responsibility and obligation. At the start of this chapter, I stated that our first task is to indicate the direction 'in which we ought to be headed', and I believe it is these considerations that can show us the way forward.

2 'Spaceship Earth': A View on the History of Earth

Yosuke Yamashiki

Earth is the only planet that is known to be able to support human life in our Galaxy. However, since 1995 when Michel Mayor (Mayor and Queloz 1995: 355–359) discovered the existence of a Jupiter-mass companion, later confirmed as a first extrasolar planet (exoplanet) orbiting around a main-sequence star by the Radial Velocity method[1], it has become clear that there are plenty of similar planets in our Galaxy. The Jupiter-mass companion 51 Pegasi b, located fifty light-years away in the constellation Pegasus, was nicknamed 'Bellerophon', and later named Dimidium. In our solar system, Jupiter's orbit is 5.2 AU[2] from the Sun, and from the standard solar system model (Kyoto Model)[3] it was considered that gas giants formed at a great distance from the host star, thus they would never be close to it (0.052 AU).

Considering the age of the host star (almost 10 GY), it is thought that 51 Pegasi b has migrated from its original position through orbital evolution, and is now classified as a new class of planets called 'hot Jupiter'.

While the discovery of exoplanets has significantly accelerated, new categories of planets have been proposed, such as 'hot Jupiter', 'hot Neptune' (Neptunian mass planets orbiting very close to the host star,) and 'Super-Earth' (rocky planets slightly larger than Earth (within ten mass of Earth)). However, since less than 0.0002%[4] of stars in our Galaxy have been surveyed for exoplanets, and since detection methods are likely limited by technical capabilities, it is possible that there are many more. There are thought to be around seventeen billion Earth-like planets in our Galaxy (NASA News 2013; Batalha et al. 2013), and within those, it is possible that a planet exists that has oceans, a temperate climate and is in the habitable zone (called the Goldilocks Zone[5]). However, detection of these small planets poses a difficult task due to their smaller size,

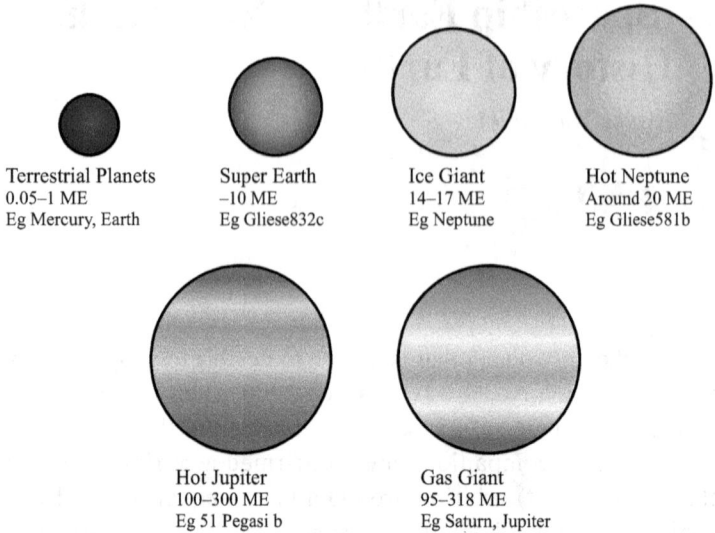

Figure 2.1: *Classification of extrasolar planets*

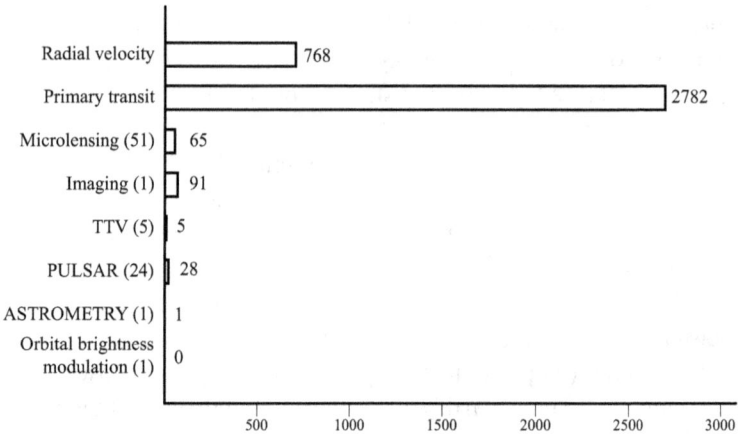

Figure 2.2 *Classification of extrasolar planets by method of detection*

Note: Classified using ExoKyoto on January 1, 2018.

in comparison to Jupiter-sized planets, and longer orbital periods. In March 2009 the Kepler space telescope launched, making a groundbreaking discovery. The planets found by Primary Transit,

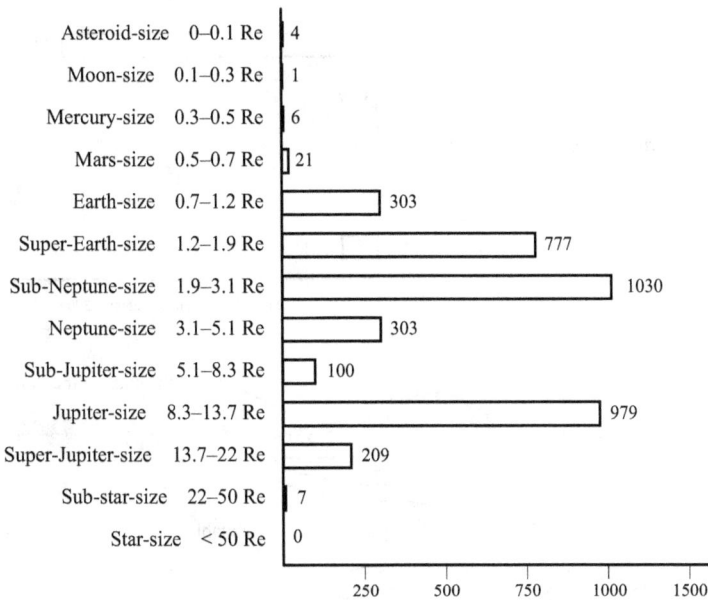

Classification	Size	Count
Asteroid-size	0–0.1 Re	4
Moon-size	0.1–0.3 Re	1
Mercury-size	0.3–0.5 Re	6
Mars-size	0.5–0.7 Re	21
Earth-size	0.7–1.2 Re	303
Super-Earth-size	1.2–1.9 Re	777
Sub-Neptune-size	1.9–3.1 Re	1030
Neptune-size	3.1–5.1 Re	303
Sub-Jupiter-size	5.1–8.3 Re	100
Jupiter-size	8.3–13.7 Re	979
Super-Jupiter-size	13.7–22 Re	209
Sub-star-size	22–50 Re	7
Star-size	< 50 Re	0

Figure 2.3: Classification of extrasolar planets by size

Notes: Re = Earth's radius; Extrasolar planet's size classification using ExoKyoto on December 30, 2017.
Sources: Exoplanetkyoto.org, Exoplanet.eu and openexoplanetcatelog.com.

including Earth-size small planets, have increased significantly, and this has become a primary method of detecting extrasolar planets.

According to classification using the ExoKyoto (exoplanetkyoto. org) database, by October 10, 2017 a total of 3,681 planets have been confirmed (including some within our solar system). Among them, 995 are within Sub-Neptune-size (1.9–3.1 R_{Earth}), 696 within Super-Earth-size (1.2–1.9 R_{Earth}) and 286 within Earth-size (0.7–1.2 R_{Earth}).

However, most of these planets are situated in horrible locations to the host star, as 3,034 among 3,540 are inside the Venus Equivalent orbit, classified using Solar Equivalent Astronomical Units as below.

Only a few of them (twenty-three) are located in the Goldilocks Zone, determined by Kopparapu et al. (2013) using ExoKyoto as below.

Even if we locate these planets, however, it is impossible to travel to even the closest ones with modern technology (Proxima Cen b is situated 'only' 4.3 light years from our solar system, so eventually, future generations could migrate to it).

Inside Venus equivalent — 3122

Venus equivalent – Earth equivalent — 99

Earth equivalent – Mars equivalent — 109

Mars equivalent – Snow line — 89

Outside of snow line — 321

500 1000 1500 2000 2500 3000 3500

Twice the mass of the Sun

A stars
5–25 L⊙
About 2AU

Habitable zone

F stars
1.5–5 L⊙
AU
Venus

G stars
0.6–1.5 L⊙
Earth
Mars

K stars
0.8–0.6 L⊙

M stars
< 0.8 L⊙
About 0.5AU
Half the mass of the Sun

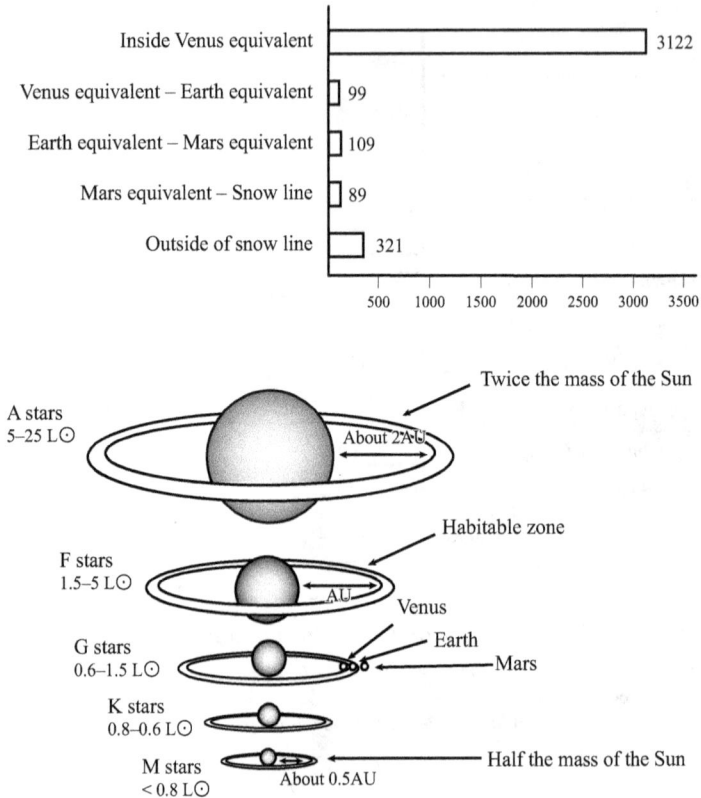

Figure 2.4: Classification of extrasolar planets by relative position
from host star using Solar Equivalent Astronomical Unit
Note: Classified using ExoKyoto on November 30, 2016.

Of course, there is also insufficient proof of the possibility that a spaceship from another planet capable of light speed will ever visit Earth. And unfortunately, to earthlings, traveling at the speed of light is only real in the movies, and there is no evidence that intelligent life has discovered warp speed navigation. For that reason, the Earth is alone on the list of 'habitable' planets for human beings. In this chapter, we look at what kind of 'planet' 'Spaceship Earth' is, and contemplate the fate of a star.

Many different types of planets orbit a star. The central star of our solar system, the Sun, is a G type star. According to nuclear

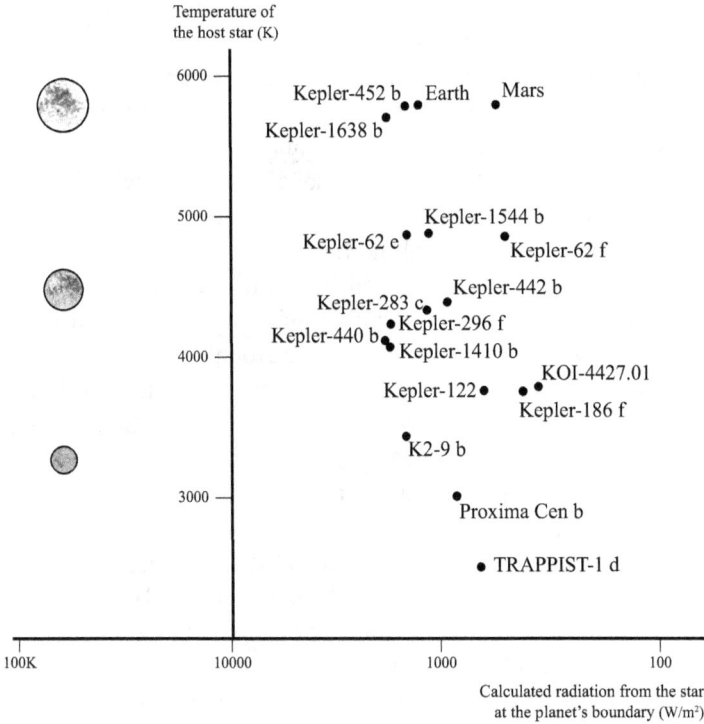

Figure 2.5: Habitable planets located in Goldilocks Zone determined by Kopparapu et al. (2013) and their estimated radiation from the host star

fusion (PP chain) theoretical values, the life expectancy of this kind of star is about ten billion years. However, a star with a solar mass of 1.4 to 2.1 times, for example, the blue-giants from a summer night sky in constellation Vega, or the extremely bright Sirius in the Canis Major constellation, are classified as A type stars. These stars have a superior CNO cycle using nuclear fusion, from the PP Chain's efficient reaction; it is calculated that these will have a lifespan of one billion years.

Furthermore, there are the B type stars, like Spica in the Virgo constellation, whose life isn't expected to last more than 100 million years. The planets that orbit these types of stars should have much shorter lifespans, unlike Earth.

On the contrary, if you look at stars that are lighter than the Sun, stars with a solar mass of 0.45 like Gliese 581, or one of the closely located stars from our solar system such as Barnard's Star, the lifespan is estimated to be 100 billion years. This means that the lifespan of the planets in their solar systems is much longer. With these types of stars, there is thought to be sufficient time for intelligent life or civilizations to develop on the planets in their solar systems. With humankind thought to have emerged around 4.5 billion years after the establishment of the solar system, it is important to consider that the birth of civilization on Earth was possible because it happened to revolve around the Sun.

It should also be noted that the current solar flare activity is relatively small (up to 10^{32} ergs) compared with other Sun-like stars, in which Maehara et al. (2012)[6] discovered much more severe stellar flares (10^{34}–10^{36} ergs). This also provides a moderate environment for our generation's developed ecosystem.

With eight planets in our solar system, why was Earth the one that formed life? It can't be said that we haven't thoroughly searched our solar system for life. It is possible that there could be life on Mars, Europa and Saturn's satellites Titan and Enceladus. But, although this is exciting, it is unlikely that these planets have any type of civilized life.

The most important factor in the creation of the diverse life on Earth is the ocean. Regarding distance from the Sun, for a planet to have oceans its position has to be further than Venus and closer than Mars. Although it is thought that Venus had oceans in the past, a runaway greenhouse effect[7] caused them all to evaporate, resulting in a loss of water in the outer atmosphere. It is speculated that at the time of its creation the Sun was only about 75% of its current brightness, and it is getting brighter. It was in this early period that there was a possibility that Venus's orbit was sufficient to create oceans.

On the other hand, if all the ice on Mars were to melt, an ocean with an average depth of thirty cm would cover it. If Mars had the same percentage of water as the Earth (weight basis) it would have oceans thousands of meters deep. Scientists believed that Mars had oceans, but lost them when the magnetic shield was diminished by the cooling of its core. Thus Venus and Mars are called Earth-like planets, and it wouldn't be too strange to think they could support life given the right conditions. One of the biggest concerns is the

	O	B	A	F	G	K	M
	10 Lacertae	Spica Rigel	Sirius Vega	Procyon	Sun Centuari	Elidani Pollux	Gliese 581 Barnard's star
	10^7 years	10^8 years	10^9 years	Lifetime	10^{10} years		10^{11} years
	CNO cycle	CNO cycle	CNO cycle	PP chain CNO cycle	PP chain	PP chain	PP chain
Surface temperature	≥ 30000 K	10000–30000 K	7500–10000 K	6000–7500 K	5200–6000 K	3700–5200 K	2000–3700 K
Mass	≥ 16 M☉	2.10–16.00 M☉	1.40–2.10 M☉	1.04–1.40 M☉	0.80–1.04 M☉	0.45–0.80 M☉	≤ 0.45–0.80 M☉
Radius	≥ 6.6 R☉	1.80–6.60 R☉	1.40–1.80 R☉	1.04–1.40 R☉	0.96–1.15 R☉	0.70–0.96 R☉	≤ 0.70–0.96 R☉
Luminosity	≥ 30000.00 L☉	25.00–30000.00 L☉	5.00–25.00 L☉	1.50–5.00 L☉	0.60–1.50 L☉	0.08–0.60 L☉	≤ 0.08–0.60 L☉

Bar chart values:

Category	Value
O	0
B	13
A	25
F	711
G	1819
K	928
M	227
Red Giant	83
Pulsar	15
WD	2

(horizontal axis: 500, 1000, 1500, 2000)

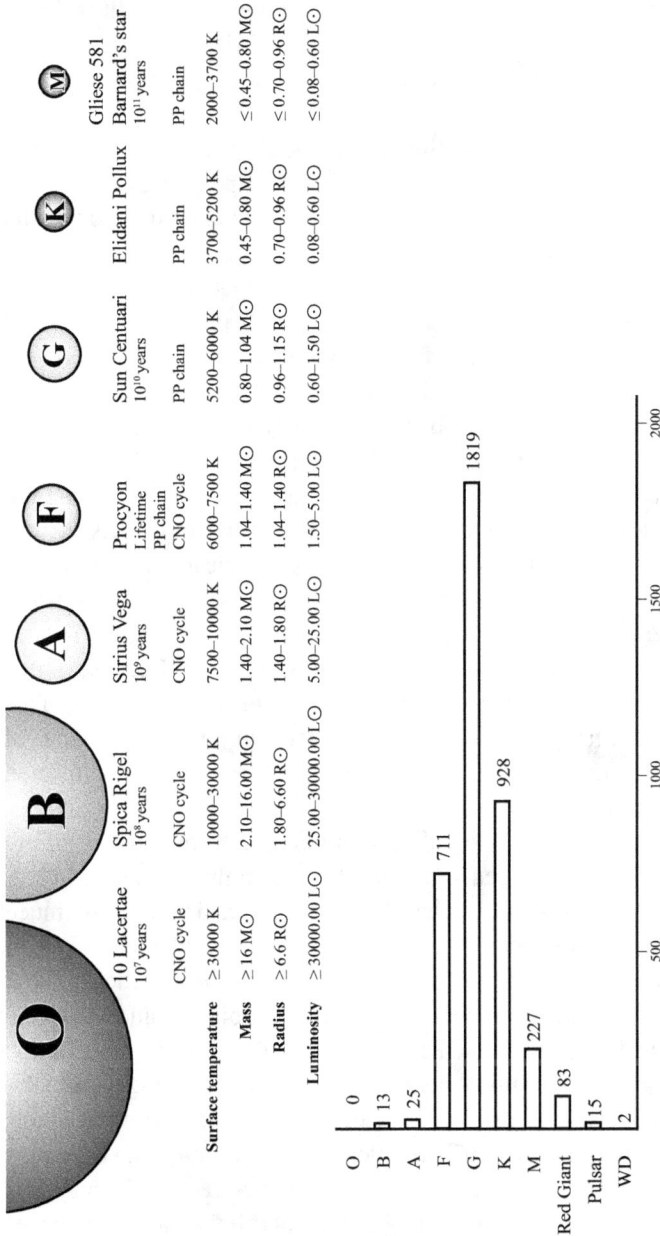

Figure 2.6: Stellar classification: Host star's spectral type for each exoplanet

question of whether a faint young sun can provide enough heat for Earth to become not completely frozen (faint young sun paradox)[8], which has been explained recently (Airapetian et al. 2016). Our active young Sun's powerful superflares (CMEs) generate nitrous oxides as powerful greenhouse gases and prevent the Earth from freezing.

Nevertheless, Earth is the only planet in the solar system that has been stable long enough to have oceans, and though oceans and continents emerged long ago, they have a deep connection to the evolution and life of 'Spaceship Earth'.

Temporarily, what if there were only oceans and no continents? Of course, humans would not have emerged, and it is not known if ocean life would be able to form a sophisticated civilization. There is only a weak chance for the development of human civilization and infrastructure, so the formation of the same type of civilization in another setting is utterly impossible. If you look at the history of Earth, continents were formed over an extremely long period of time, and those continents repositioned, connected and divided repeatedly.

Given this situation, why were continents able to form at all? The main cause is the formation and elevation of granite. Lava and water together form granite, which has a smaller specific weight ($2.7g/cm^3$) compared to other rocks like basalt (3.0–3.3 g/cm^3). Because of this buoyancy, it is said that continents can form from their core. This reaction with water can, of course, only occur with the existence of oceans, so thanks to the oceans, light rock was able to be formed into continents.

Looking at Earth's 4.6 billion year history, the blossoming of life activity primarily occurred from the Precambrian Supereon Era. From 4.6 billion years ago, the freeze was over, and this era continued until 540 million years ago. After that were the Phanerozoic Eon, Paleozoic Era, Mesozoic Era and Cenozoic Era. During each of these eras a large extinction occurred. The Paleozoic Era and Mesozoic Era were separated by the Permean-Triassic (P-Tr) Mass Extinction Events and the Oceanic Anoxic Event, where it is thought that 90% of life on Earth was destroyed. Also, the Mesozoic Era and Cenozoic Era were separated by the extinction of the dinosaurs, and the end of the Cretaceous Era was marked by the K-T (Cretaceous-Tertiary) and K-Pg (Cretaceous-Paleogene). The cause of this was large meteorites the size of Mount Everest crashing into the Yukon peninsula. This has been verified through various investigations.

Other than those large extinctions, there is the Triassic-Jurassic (Tr-J), Late Devonian (Late D) and Ordovician-Silurian (O-S) events. In the O-S event, the Earth's temperature suddenly dropped. After that, it is not clear what caused the last three extinction events (P-Tr, Tr-J, Late D). In the P-Tr event, by means of the birth of the Pangea continent, there was a large outbreak of volcanic activity caused by the flooding of basalt in Siberia. However, as for the other two events, there are no clear causes.

Humans emerged seven million years ago; civilization was developed about 20,000 years ago after the last glacial period. Earth's history is extremely short. After the emergence of humans, the Toba volcanic eruption occurred, which rapidly reduced the population of India. From this perspective, in order to form a civilization, historical inspection is needed.

Thinking of the future of humanity, and the future of Spaceship Earth, we need to study in depth the past extinction events in case they happen again, and take this possibility seriously. We need to study these in order to avoid future catastrophes.

3 Lessons from the History of Life

Masao Mitsuyama

Introduction

The Earth is full of life. It is no exaggeration to say that we humans live by being supported by the lives of other creatures. We consume large amounts of fish and shellfish from rivers and seas and more recently from aquaculture. We also consume domesticated animals as sources of protein. We breed plants with high nutritional value, grow them on the greatest scale possible and utilize them as sources of carbohydrates, vitamins and vegetable protein. At the same time, we use animals and plants (especially those with flowers) as pets and ornaments for the purpose of enriching our lives. Clearly, human beings are not the only living organisms on Earth. It is time to consider the history of life from a somewhat biological perspective, based on the fact that not only human beings but also many other versatile living organisms live on Earth, interact with each other and are part of Earth. Furthermore, what does it mean to preserve life, and what do living organisms mean to the Earth and nature?

What is life?

To begin with, let's consider what life is. Many living and non-living things co-exist on Earth. It is hard to define 'life', surprisingly, although it seems a simple term. Generally, if a creature meets the following two conditions, it is called a 'living thing': (1) it has the genome defining its genetic background and has a self-propagation ability (self-replication ability) and (2) it metabolizes and produces ATP (adenosine triphosphate), which is said to be the energy for biological activity. Therefore, human beings and bacteria are both 'living organisms' (living things) on the Earth. Generally, viruses that cause infection are categorized as microorganisms. However, when examining whether or not they are living organisms, they do not fit the criteria and are simply considered to be infectious particles of non-living things. If viruses are observed biologically, they meet condition (1) mentioned above, but do not meet condition (2). Furthermore, although both bacteria and viruses are microorganisms, bacteria, which are regarded as living organisms, are equipped with a series of functions from genome replication to the production of constituent protein by themselves (transcription from nucleic acid and genomic DNA to messenger RNA and the synthetic pathway of necessary proteins). On the other hand, viruses can only hijack and use the process of cells they infected. Therefore, although viruses can multiply and reproduce themselves, they are non-living things.

Furthermore, when a living organism dies, it is no longer considered to be a 'living thing', because living activities are no longer observed in the dead cells and body. Life is a concept where living things meeting the above conditions (1) and (2) are actively and independently preserving themselves (or their species) and multiplying under a certain order. Human beings, higher forms of being, have developed elaborate languages and spirituality and acquire very high levels of cognitive function while meeting these conditions.

The most important factor among those constituting living organisms is the nucleic acid genome and functional protein. Protein is a polypeptide of amino acids. However, it is a matter of speculation as to how and where amino acids appeared on the Earth. Four-point-six billion years ago, when the Earth is said to have been formed, the Earth's primeval atmosphere mainly consisted of helium and nitrogen, and no amino acid existed under the high temperatures and pressures. Subsequently, along with the lowering of temperatures, ammonia and carbon dioxide were released from the Earth's crust and all the elements of amino acid, nitrogen, carbon, hydrogen and oxygen, were then present. In 1953, it was proved that if the combined gas of methane, ammonia and oxygen is charged with electricity, organic matter including acetic acid and amino acids such as glycine and alanine are produced. It is considered that the fundamental factors for life came into existence in the process of the evolutionary history of the Earth as an astral body. Aside from this, based on research on meteorites and planets, it is thought that the origin of life on Earth might have come from extra-terrestrial sources. However, this hypothesis is not generally accepted.

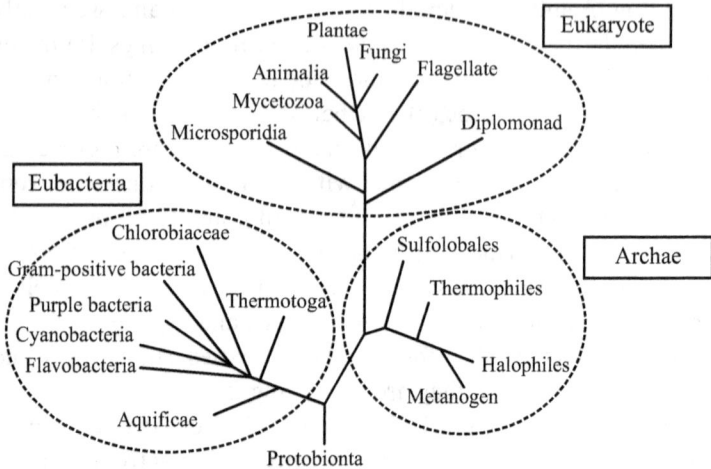

Figure 3.1: Phylogenetic tree of life: Three-domain system

Evolution of living organisms

In current evolutionary biology, the phylogeny of every species of organism was recently established by a detailed molecular biology analysis of the existing species of organisms and, in particular, the comparison of genome sequences. The organisms on Earth can be classified into three domains of the highest biological classification. All organisms can be largely classified into eukaryotes, which include all plants, animals and fungi, and prokaryotes, which consist of bacteria. Prokaryotes can be further classified into eubacterium and archae (Figure 3.1). As viruses are non-living things, they are not included in these three domains. In conjunction with the above, prokaryotes are characterized as having no nuclear envelope or nucleolus and no subcellular organelle, such as a mitochondria or Golgi apparatus. Given the general impression that more than half of the living organisms on the Earth are plants or animals, it may seem strange to divide the living world into two domains of tiny microorganisms. However, according to some trial calculations, astonishingly there are 415–615×10^{28} individual microorganisms on the Earth (the number of human beings is 737×10^7 individuals at present), and more than half of them exist below the bottom of the sea. Archaea are bacteria that exist under high temperatures and pressures, and at first glance, they appear to have adapted to the environment at an early stage of the Earth's formation. However, judging from their genome composition and protein biosynthetic pathway, they are considered to be closer to higher animals, such as human beings, than general eubacteria, such as coliforms. It is considered that primordial cells did not evolve into archaea and then into eubacteria, but instead into two branches: one containing eubacteria and another archaeas and eukaryotes.

What is interesting is that even though they are subcellular organelle, mitochondria that exist in the cells of higher animals play an indispensable role in terms of energy metabolism as they have lipid bilayer membranes and mitochondrial DNA independent from genomic DNA. There is even a theory that after our progenitor cells emerged in ancient times, the already existent bacteria parasitized them and remained there as subcellular organelle constituting eukaryote cells (symbiotic evolution theory). From the perspective of Darwinian evolution theory, it is considered that primordial cells were formed as aggregations of various organic substances generated on the Earth and from which eubacteria and archaea and then eukaryote cells evolved. They continued to evolve in each domain and as a result the living organisms that we recognize today were formed (Figure 3.1).

Evolution of human beings

Human beings are the most evolved species of living organism. We, contemporary human beings, belong to the genus *Homo sapiens*, classified into genus *Homo*, hominidae, primate, mammalian and chordate. The genus *Homo*, which is older than *Homo sapiens*, had *Homo neanderthalensis* (about 250,000 years ago), *Homo erectus* (the Java man and Peking man, formally known as '*Pithecanthropus erectus*' about 1.9 million–150,000 years ago), and *Homo habilis* (about two million years ago). Before then was the *Australopithecus* species in hominidae, and this fossil hominid was considered to exist from four million to two million years ago. *Australopithecus*, whose skulls were found in South Africa, are estimated to have been 120–140 cm in height, with small brains of about 500 ml, and are thought to have walked upright on two legs. It is said that this archaic human group is the mammal intermediate between man and ape, and *Homo habilis* is said to have evolved from this.

Therefore, where were contemporary humans (*Homo sapiens*) born? Regarding *Homo sapiens*, which came into existence 250,000 years ago (*sapiens* means wisdom and brightness in Latin), there are two theories. One is an African single origin theory where *Homo sapiens* propagated and expanded from Africa to the world, and another is multiregional evolution theory where humans who existed earlier than contemporary humans and *Homo neanderthalensis* evolved into contemporary humans in various areas of Africa, Asia and Europe.

There are limitations to archaeological verification based on fossils, because the research specimens are limited as are the locations producing fossils due to environmental changes that occurred in ancient times. Therefore, the geographical distribution of human genomes is very useful. There are maternally derived mitochondrial DNA (mtDNA) in the mitochondria existing in our cells. Different from genomic DNA, the hereditary change of mtDNA does not depend upon recombination; therefore, if there is any change in sequence, it can be considered to be due to the mutation that occurred probabilistically in a certain period of time. Cann et al. (1987) of the US selected 147 subjects of various races and examined their mtDNA base sequences, and as a result, obtained two dendrograms. One consisted of only Africans and the other consisted of Africans and all other races. This result suggested that the most common, closest maternal ancestral link to contemporary humans was in Africa, and they named her 'Mitochondria Eve'. Whereas mtDNA is maternally derived, the analysis of the Y chromosome, a sex chromosome that only men possess, was conducted. By the dendrogram obtained from the comparative analysis of the mutations observed in the sequences in the male-specific region of the Y chromosome (MSY), the paternal ancestor common to contemporary humans was traced back to Africa 50,000 years ago; he was named 'Y-chromosomal Adam' (Thompson and Pitchard et al. 2000). Including the above results, the idea that the ancestors common to contemporary humans were born in Africa and subsequently spread all over the world is dominant.

The origin of human beings can be presumed not only from the analyses of fossils and the genes of human beings, but also from the genes of bacteria housed by humans. *Helicobacter pylori* are bacteria that live in human stomachs, and it was recently established that they cause chronic gastritis and gastric ulcers. As the bacteria's generation time (time for division and proliferation to occur) is short, various mutations can be recognized. It is possible to analyze, by computer, the time period over which the changes occurred by analyzing the genome sequences of *Helicobacter pylori* collected from all around the world. The diversity of the genes of *Helicobacter pylori* collected from East Africa, in particular, was small, suggesting that *Helicobacter pylori* spread from Africa about 60,000 years ago (Linz and Balloux et al. 2007). In this manner, the idea that human beings originated in Africa is also supported by the gene analysis of bacteria that specifically coexist with humans.

It goes without saying that we, contemporary humans, are positioned at the top of the primate hierarchy. The characteristics that locate human beings above other primates are (1) walking upright on two legs and (2) having large brains. We learned to use fire about 500,000 years ago and by cooking with fire we effectively took nourishment from food that was difficult to eat raw. This made it possible to preserve food for longer periods of time. This phenomenon is not observed in any other animal. We invented earthenware 10,000 years ago and the wheel 5,000 years ago, making it possible to store and carry various materials. We invented paper around 100 AD, and the mariner's compass around 1000 AD, making it possible to carry out long-range transport by following preserved records and accurate sea routes, and as a result, civilization and culture rapidly developed. Although the evolution of living organisms is generally explained by the evolution of genomes, the evolution and development of human life by invention and discovery do not significantly relate to this process. They may, however, have been caused by the work of the human brain, which is the most relatively enlarged in the animal world, in other words, by learning, memory and tradition.

Even though human beings are said to appear similar to chimpanzees, there is a significant difference in brain activity. However, the difference in genomic DNA between human beings and chimpanzees is only 1.6%, and 98.4% of their genomic DNA is the same. This suggests that the quality of living organisms, including social activity, is not defined solely by the evolution of genomes, but also by the volume and development of the brain and its functions that bring about the ability to adapt to the environment and vitality.

It is known that the volume of the brain differs according to animal species. The brain weight of larger animals is heavier and thus does not afford simple comparisons. Figure 3.2 shows the average brain weight of animals adjusted according to a common body weight of

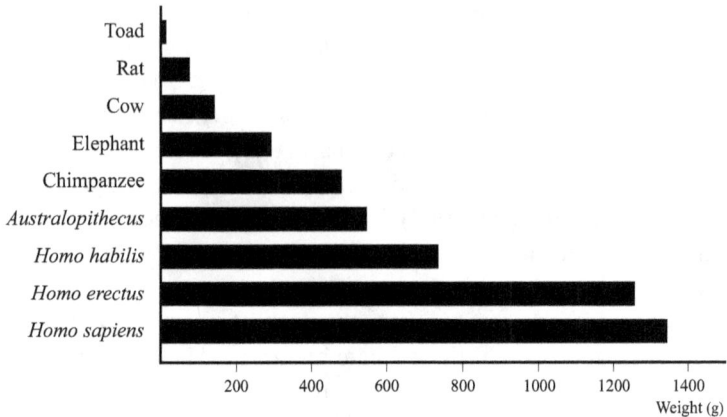

Figure 3.2: The relative comparison of brain volume of various animals

Note: The average brain weight of animals adjusted according to a common body weight of 60 kg of each animal.
Source: Adopted and revised from Newton (2008).

sixty kg. Generally, higher animals show higher figures, and the brain volume drastically increased from *Homo erectus* onwards. However, according to this simple calculation, in general the brain volume of human beings is 1/40 of their body weight while that of birds is 1/12, which is larger than the figure for human beings. Therefore, the relative value of brain weight as an intellectual factor is not always practical. According to the encephalization quotient (EQ), calculated by the formula of (constant) × (brain weight) ÷ (body weight)$^{2/3}$, the EQ of human beings is 7.5, that of chimpanzees and squirrel monkeys is 2.3, those of rhesus macaques and gibbons are 2.1 and 2.0, respectively (showing that the EQs of monkeys are almost the same), and those of elephants and cats are 1.3 and 1.0, respectively. The brain volume of human beings rapidly increased from around *Homo erectus pekinensis*, which may account for the unique intellectual activities of human beings.

Diffusion and expansion of human beings

It is generally believed that the ancestors of humans were born in Africa and spread to Europe and Asia about 100,000 years ago.

According to a recent theory (Toba catastrophe theory), the Toba volcano on Sumatra produced a large eruption about 75,000 years ago, rapidly cooling the Earth due to the sun-blocking effect of the large amount of volcanic dust, and this continued for 6,000 years. The rapid change in weather caused by Toba's massive eruption annihilated the majority of human species; however, *Homo sapiens* survived. It is thought that the number of *Homo sapiens* decreased to 10,000, and owing to the bottleneck effect, generated a human group with relatively less genetic diversity (with relatively uniform genetics).

Human beings continued to expand in number, and according to the 'State of World Population 2011' by the UN, the world population was estimated to reach over seven billion in October 2011. Although it is almost impossible to determine demographic dynamics of the past, it has been estimated by various data that the population was about 150 million at around 400 BC, 170 million around 1 AD and 200 million around 600 AD (Hayami 2012). This gradual population growth continued to around 1500 AD (about 430 million), and during this period, it took about 1,000 years for the population of the world to double.

After that time population growth gradually accelerated, and it only took about 250 years for the population to double from 400 million to 800 million, and only about 100 years for it to double again to 1.6 billion. In the twentieth century, its doubling time was shortened to fifty years (Figure 3.3).

Since the eighteenth century, human population growth has been drastic, both in terms of rate of increase and scale, as shown in the arithmetic graph where the increase is expressed as an almost vertical line. It is necessary to recognize that this phenomenon occurred in parallel with the start of human history. Surprisingly, the current population of the world is equivalent to one-fifth of the total population that existed on the Earth over the previous 6,000 years.

Many of the problems we are facing in terms of poor distribution and shortages of food, water and energy are mainly caused by the increased growth rate in the human population. Many countries and ethnic groups work collectively to create access to food, water and energy, and as a result, problems of conflicts and refugees emerge. At the same time, it should be noted that human beings' positive aspects, based on human intelligence, have largely contributed to this rapid population growth since the Industrial Revolution. Food production to cater to larger populations has been markedly increased by irrigation,

Population
(Billion)

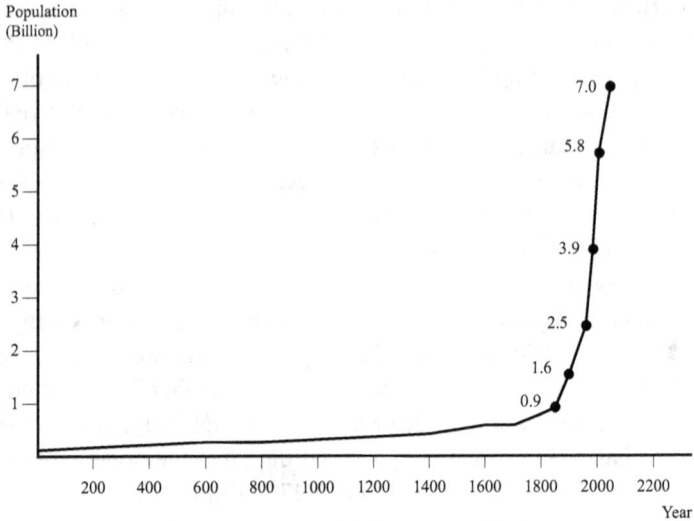

Figure 3.3: Speculated changes in the world population
Source: Adopted and modified from Hayami (2012).

the expansion of arable land by land improvement and the development of large-scale livestock farming. The advancement of medicine and the spread of public health promotion have also largely contributed to lower the infant mortality rates. The average life expectancy in Japan is eighty-four years old, but it was only forty-three in the Taisho period (1912–1926). The clear decline in the infant mortality rate contributes to the lengthening of average life expectancy more than the decline in the adult mortality rate. The infant mortality rate is now only 0.2% in Japan. We should remember, however, that countries like Sierra Leone still experience an extremely high infant mortality rate, at 11.9%.

Conflict between living organisms and the environment

The number of human beings has drastically expanded in the past 200 to 300 years and the current global environment is becoming precarious on a number of fronts that impact the human population. Overcoming these problems is thus a particularly pressing issue. Each person living on Earth should comprehensively consider the purpose of human life, what they want while living and whether or not they exist only to maintain their life. Looking at it from the perspective of population change, it seems that only humans have reached the peak of prosperity.

What is happening to species other than human beings? Various living organisms have been living together with humans on the Earth, and the global environment that facilitates this has ensured biological diversity. However, there are many species that find it difficult to live in the natural world today due to overexploitation for commercial trade or drastic change to their habitats. This environmental change is not only caused by natural ecological changes, but is primarily due to human activities such as deforestation, changes to waterways and the use of pesticides, which leads to a decrease in the number of small animals and insects that provide food for other wild species. According to the 2012 Red List issued by the Ministry of Environment, forty-four animal species and sixty-six plants have become extinct in Japan. Seven mammals, including the Japanese otter and Japanese wolf, have become extinct and twenty-four species, including the famous Iriomote wild cat, Amami rabbit and Japanese sea lion, have been designated as endangered species.

The crested ibis, known as Toki (*Nipponia nippon*), was a very common species inhabiting the Tohoku district and the edge of the Sea of Japan in the nineteenth century. However, the number of this species drastically decreased due to overhunting for its feathers and meat, and it has not been seen in the wild since the 1960s. The last Japanese-born crested ibis, named Kin, died in 2003, marking the end of the species. Following this, artificial breeding using Chinese-born crested ibis was conducted and since 2008, more than 100 of the species have been released in Japan. It is unknown whether efforts such as artificial breeding will successfully reproduce the ensemble of crested ibis that once dyed the sky red, a scene that used to be witnessed in many local districts in Japan. Even though it may be based on regret for wrongdoing caused by self-interest, why do human beings strive to preserve and regenerate crested ibis among other animals? It may be necessary to consider this in relation to the preservation of the global environment in the future. It goes without saying that, once a species has become extinct, or even if it is endangered, tremendous effort and expense would be required to maintain a sustainable population in a natural environment. We should clearly recognize that when we are able to do so, the coexistence of human beings with other organisms should be harmoniously maintained.

What can history teach us?

As noted in 'What is life?' at the beginning of this chapter, the most primitive living organisms existing on Earth are bacteria. It is presumed that the number of bacteria is 10^{21} times that of the population of human beings, who have drastically expanded their numbers. Humans firmly believe that 'bacteria' is synonymous with filthiness and that they are nothing more than entities that cause infectious disease. Among the eubacteria, presumed to have more than 500,000 kinds including those that cannot even be cultivated, only 200 types can cause infectious disease in human beings and animals. One thousand types and 10^{13}–10^{14} individuals of normal bacteria exist in our intestines, and many play important roles including in the production of vitamins, maintenance of intestinal homeostasis, prohibition of the settlement of non-persistent pathogenic germs and in the development of immune response. In short, it is not an exaggeration to say that at least some degree of the life activities of human beings, the living organisms that are most highly evolved, are made possible by coexistence with the lowest form of living organism. Moreover, fermented foods, including *miso* (soya bean paste), *shoyu* (soy sauce), *natto* (fermented soya beans), wine and beer, are produced by microorganisms with particular fermentation metabolic pathways that human beings do not have. Different from autotrophs that can produce carbohydrates for energy sources from carbon dioxide in the air, the human body is heterotrophic, unable to produce energy unless it takes in external nourishment. In other words, human beings survive by taking in living organisms other than human beings and the lower-order living organisms that these organisms take in. When we consider the circle of life, as stated above, we recognize that the existence of human beings and the preservation of a global environment cannot be for human beings alone, and therefore, we should bear in mind the preservation of all life on Earth.

4 What Can We Learn from 'History'?

Takura Izumi

Introduction

Let me start with my fields of specialization – archeology and prehistory. In contrast to 'real' or 'practical sciences' such as engineering, medicine and economics, which are of direct use to society, archeology and prehistory have been called 'empty sciences' that make no such contribution. The self-image of these fields is that they are like something relegated to the back of an old museum.

This insularity inevitably started to change with the disturbance of archeological sites and environmental destruction associated with large-scale development in the era of high-level growth from the 1960s onwards. The Convention Concerning the Protection of the World Cultural and Natural Heritage (World Heritage Convention) adopted at the UNESCO General Conference in 1972 was a new global-level initiative promoted in light of such circumstances. In Europe, a movement questioning the contribution of archeology to society began in around the 1970s, based on excavation surveys and site preservation activities associated with development. In Japan, however, concern regarding the contribution of archaeology to society was slower to emerge. This was in spite of the fact that Japan has one of the world's most thriving traditions for official excavation surveys associated with development, and in spite of the advancement of excavation technology by the authorities responsible for archeology and cultural heritage.

It was not until the Great East Japan Earthquake of March 11, 2011, that Japanese archeologists started to seriously contemplate their social responsibility. The field of earthquake archeology, or 'archaeoseismology', had previously been proposed by Akira

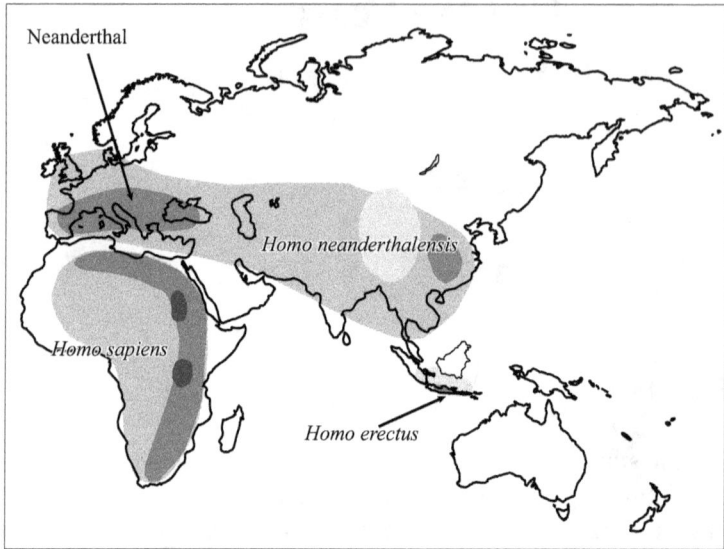

Figure 4.1: Distribution map of hominoids 100,000 years ago
Source: Baba ed. (2012).

Sangawa (1992), but was too late to be of any use when the Great East Japan Earthquake struck. Although data on disasters had been accumulated to some degree in archeological studies and excavation surveys, there was little momentum toward linking these findings to research or policy on disaster prevention, and hardly any interaction with disaster-related organizations. Since the disaster, academic societies concerned with archeology and cultural heritage have combined their resources to cooperate in the reconstruction effort through surveys of disaster-affected areas and repairs to and reconstruction of cultural heritage, etc. They have been gathering data on disaster damage and facilitated public information activities aimed at enhancing disaster prevention awareness. This marked the beginning of the social contribution of archeology in earnest.

What I found particularly shocking in connection with the disaster was the Onkalo high-level radioactive waste repository in Finland. Here, plutonium with a half-life of 24,000 years is to be entombed deep underground for 100,000 years. To imagine what sort of changes could occur over the space of 100,000 years, we can only refer to what things were like 100,000 years ago. At that time, there was no trace of modern humans in the northern Eurasian continent, the Americas or Australia, to say nothing of the Japanese archipelago; the forebears of today's human beings (*Homo sapiens*) were still evolving in Africa. Archaic humans inhabited an area stretching from the Middle East to the European continent, and it is thought that descendants of *Homo erectus* could have inhabited part of the Asian continent (Baba ed. 2012) (Figure 4.1). In other words, 100,000 years ago was a time when a human race distinct from that of today played the leading role on this planet. It is impossible for people today to predict and take responsibility for a time 100,000 years into the future, when the human race as we know it may not even exist.

Diachronic study

Historical fact

In the second half of the twentieth century, and particularly since the 1980s, historical studies were subjected to harsh criticism amid the debate on 'Postmodernism'. This revolved around the claim that 'It is an illusion that historical studies convey facts objectively. "History" is actually nothing more than a series of "statements"... It is just a story' (Ohto 2012). This raises doubts about what constitutes historical 'fact', and what objectivity is; even lists of dull and uninteresting historical 'facts' as condensed forms of historical narrative are seen as nothing more than illusion. This theory was welcomed as constructive criticism by some, and received as masochism by others. Nevertheless, it is certain that some aspects of historical studies required considerable introspection. Even 'facts' of the past that appear objective at first glance are the result of choices based on time (nature), by artifice, or made by researchers; even if they form part of the 'truth', they are sometimes divorced from this 'truth'. When trying to view things diachronically, neither human history nor natural history can escape from the spell of such 'selective results'. It is impossible to discuss various past 'facts' without being aware of this point. However, even while taking this as a premise, using these 'facts' (despite their limitations) to investigate the history of various social issues that face contemporary society has meaning in terms of solving the problems confronting us today.

The scope of historical studies

According to one view, science is supposed to deal with general and universal questions, and its explanations must be nomothetic (Ohto 2012). Geology, paleontology, geophysics and astrophysics, as 'sciences' that also deal with time, have generally been thought of as based on principles differing from those of historical studies, which idiographically researches single-event, particular problems concerning the timespan of the human race. But is it really impossible to link the two together?

Historical studies are mainly based on documentary and written archival materials, and have developed with their focus on the history of institutions, states, societies and economies since the existence of calendars. For this reason, human history has been separated from natural history, even though the human race arose amid the natural history-based chronology of the Earth (Ono 2000). Akira Ono criticizes this, saying that the genesis and evolution of the human race cannot be discussed without taking the environment and the other species surrounding humankind into account, in that it emerged amid the process of natural history. Ono asserts that history is inherently concerned with everything since the beginning of the universe, based on the rationale of the hierarchical structure of biosystems, or on the locomotive morphology of matter and hierarchical aspects of nature based on dialectical materialism. In the case of prehistory and archeology, for example, this position was taken by V. Gordon Childe, a professor at London University. In his 1936 work *Man Makes Himself* (translated into Japanese in 1951), he attempted to bring archeology (historical studies) within the scope of sciences by starting from the 'history' before the appearance of the human race, and stressing the relationship between the human race and nature. The idea of tracing history back to before the appearance of the human race is not the sole domain of researchers in historical materialism. Recently, historian David Christian wrote *Maps of Time: An Introduction to Big History* (2004), in which he describes 'history' from the creation of the universe until the present day. His research forms part of a recent trend toward 'global history' (Mizushima 2010). In that it deals with a time before the human race existed, this 'big history' includes natural history, providing a major turning point in the relationship with environmental history.

History from the environment and ecology

Fernand Braudel, a leading historian in the Annales School, spent most of World War II in a prisoner of war camp where he wrote his major work *The Mediterranean* (1966 [1949]). More or less a contemporary of Childe, though with a different background, he exerted a huge impact on historical studies in the second half of the twentieth century. According to Braudel, time is the most important aspect of history. In *The Mediterranean*, he pictures history as a tri-level structure of time. The first level is 'unmoving' history, consisting of topography and environments that hardly ever change and form the basis of history. The second is the 'gentle rhythm' history of human groups that form economies, states and systems. Finally, the history of 'rising waves' deals with short periods of time involving everyday politics and individuals. In particular, Braudel's emphasis on the link between topography, the environment and 'history' as elements of 'unmoving' history sowed the seeds of new historical studies after the war.

Although this new approach to historical studies was not an instant hit with Japanese historians, it is currently gaining strength, aided by the spread of 'global history' (to be discussed below).

A Japanese researcher, ecological anthropologist Tadao Umesao, also incorporated the environment into his view of history. Umesao's theory (1957) can be viewed as an attempt to understand the parallel evolution of lifestyles in communities on the Eurasian continent by using the transition theory in ecological studies as a model. In so doing, Umesao took the position not of the conventional empty science-type theory that emphasizes cultural lineages, but that of functional theory, which examines the lifestyles of communities as the carriers of culture. This position concurs with the views of history and culture held by Julian Steward and other American anthropologists from 1950 onwards, who raised the banner of 'cultural ecology', stressed the importance of environmental conditions in cultural evolution and attempted to systematically analyze the relationship between cultural and social structures and the natural environment. It also finds resonance with Braudel's view of history.

Temporal-spatial reach of global history and HSS

A new tide of historical studies with their focus on world history is now on the rise, as a systematic extension of the new trend in historical studies mentioned above. It is called 'global history'. The focus of this movement is the US, where historical studies are not necessarily accorded high status. The World History Association, founded in 1982, is said to have 'played a significant role in its diffusion'.

In this chapter, the aim is not to raise questions about global history itself, but to leave the fine details to the work of Tsukasa Mizushima (2010), and to introduce only those points connected with this discussion. According to Mizushima (2010: 2–4), global history has the following five characteristics.

1. The length of its timespan. It not only covers the period between the birth of the prehistoric human race and the present, but also goes back to the beginning of the universe in some cases.
2. The breadth of its geographical range. It addresses structures and movements of whole continents and ocean regions.
3. Relativization of the European world, or of the Eurocentric view of history.
4. Emphasis on interaction and reciprocal influence between different regions.
5. Besides the conventional themes of war, politics, economic activity, religion and culture, it also deals with other topics such as epidemics, the environment, population and living standards as central issues.

The characteristics of global history as outlined by Mizushima are extremely close to the historical aspects of domestic and overseas social issues addressed by Human Survivability Studies (HSS). Global history may be considered sufficiently useful as a diachronic research method in HSS.

In addition, it should be stressed that diachronic research in the form of HSS also depends on the perspective of synchronic (simultaneous) world history. Not only since globalization, but also in both historical and prehistoric eras, making historical comparisons of the whole planet seen from a 'simultaneous' perspective must have stronger contemporary significance as research on human culture and society in response to environmental change on a global scale.

Social demands of diachronic research

Although diachronic research and practice in HSS are currently under development and nothing concrete can yet be stated, the social demands of historical perspectives related to climate change as one environmental problem will now be introduced to demonstrate some relevant facts. Mizushima (2001) noted that, according to archaeological anthropologist J.D. Gunn (2000), the 1995 and 1997 IPCC reports stated that there was rising concern over the strong effects of global-scale climate change on people and communities. The latter report, in particular, introduced archaeological research by citing historical cases of extreme weather when states and people were thrust into a harsher environment than that of today, to provided the key to future research. In other words, as stated in the Introduction of this book, 'the frequency and scale of earthquakes are known to follow a pattern of power distribution'; intervals between disasters increase in proportion to their scale, and actual cases of responses to massive disasters by human society can only be seen through a historical perspective. In a recent conference on global-scale climate change, Gunn stated that archeology will be able to provide important data in future. This shows that archeology and historical studies can, by promoting research on social change in past times of extreme climate change and the realities thereof, contribute to solving social problems arising as a result of climate change (global warming), now a major social issue.

In other words, it seems that HSS has much to learn from 'history'.

Climate change in historical studies and archeology

As the problem of global warming became evident, a massive volume of data on historical climate change was provided. Much of the data resulted from stable isotope analysis of oxygen and carbon in boring samples collected all over the world, as well as from pollen analysis, mineral analysis, tree-ring analysis and other natural science research. More than anything, chronological comparison holds the key to matching climate change events with historical and archeological data. Thanks to improvements in the precision of radiocarbon dating and the recent spread of tree-ring, ice-core and clay varve dating, dating has become more accurate. Comparisons with the dating of features and artifacts found in archaeological sites has progressed, and broadly speaking, it is now possible to match events proven by analysis in natural sciences to those in historical studies, and to events revealed by archeological excavation surveys.

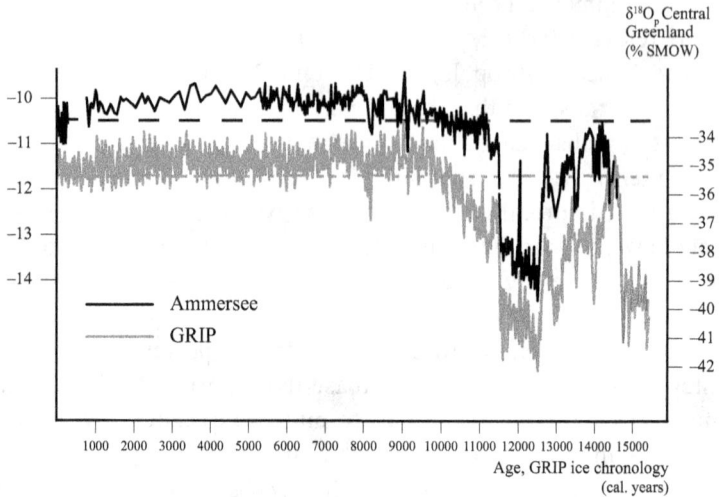

Figure 4.2: Younger Dryas Period
Source: Burroughs (2005).

Long-term climate change: Periods of maximum cold

In the last glacial period (ca. 15,000 years BP), there was a period of warming known as the 'Bølling-Allerød interstadial'. Evidence of this warming can be found all over the world. In analysis of ice cores collected from the Gregoriev Ice Cap in the Tian Shan mountains of Central Asia by a Japanese survey team, it was revealed that during this period there were no glaciers and that soil was formed, and it could even have been warmer then than it is today (Kubota ed. 2012). Archeological data show that a new culture called the Natufian emerged in the Middle East during this period. This culture was characterized by the use of storage pits for wild wheat and barley, and highly sedentary settlements with dwellings and graves created in residential communities. In Japan, too, people of the Jomon linear relief pottery culture (the earliest Jomon culture) stored acorns, chestnuts, walnuts and other nuts in storage pits, developed pottery for cooking and processing foods and built settlements consisting of pit dwellings, storage pits, hearths and graves. First established on Kyushu Island, where the warmer climate suited this new lifestyle = culture, it then spread to the northern end of Honshu Island during the period of warming.

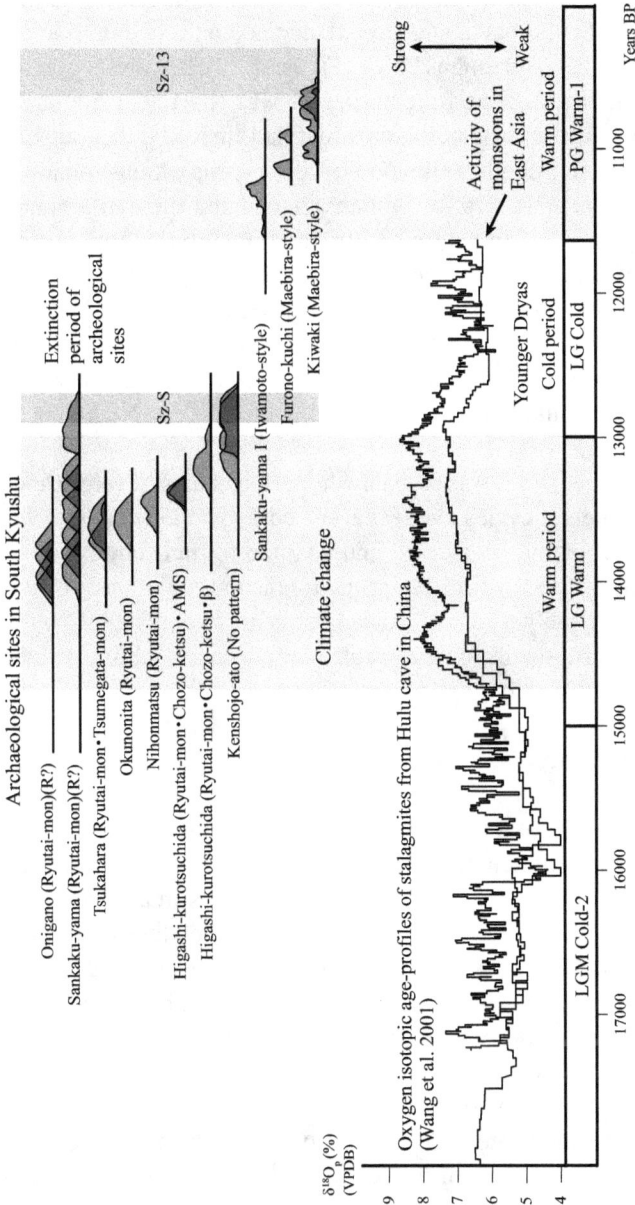

Figure 4.3: Extinction of archaeological sites at Younger Dryas Period in South Kyushu Island

Source: Kudo (2012).

The boon of this warm climate came to an end 12,900 years BP, when the whole globe was suddenly hit by a new cold wave. This corresponds to a cold period called the Younger Dryas that occurred in Scandinavia and Greenland and continued until about 11,600 years BP (Burroughs 2005) (Figure 4.2). The cooling started quickly and the decline in temperature was sharp, causing mass extinctions of the large herbivores that had thrived during the ice age such as mammoths, elk and buffalo. In this, though small in scale, some researchers see climate change and the extinction of species caused by an asteroid collision. During the cold period, sites of the Jomon linear relief pottery culture that had thrived until then were abandoned; there are hardly any sites from the Younger Dryas in southern Kyushu (Kudo 2012) (Figure 4.3). In the Middle East, conversely, the inferior environment promoted the cultivation of wild wheat and barley and led in turn to the establishment of crop farming.

The Younger Dryas was probably the period of most severe change in environmental cycles ever experienced by *Homo sapiens*. Even then, human beings dispersed and adapted to their environment, responding differently to different environmental conditions. Today's situation, arguably a case of 'maladjustment' to climate change in the present era, seems to be a problem of our monoculturally fixed contemporary society.

Pinpoint event: AD 536 – The 'Year without Summer'

In November 2013, a warrior wearing iron armor was discovered under thickly sedimented volcanic ash at the Kanai Higashiura Site in Shibukawa City, Gumma Prefecture. Perhaps in an attempt to pacify an eruption of Mount Haruna, he had fallen face down with both of his knees on the ground, the toes of both feet flexed and his head pointing west. The pyroclastic flow that must have swallowed him up instantaneously was caused by a massive phreatomagmatic eruption of Mount Haruna some time between the end of the fifth and the beginning of the sixth century. Several years after the eruption, the evacuated villagers returned to the area and made new rice paddies on top of the sedimented volcanic ash. But about thirty years later there was another massive eruption of magma from Mount Haruna, burying the paddy fields under a thick layer of volcanic ejecta. No more paddy fields were created after this.

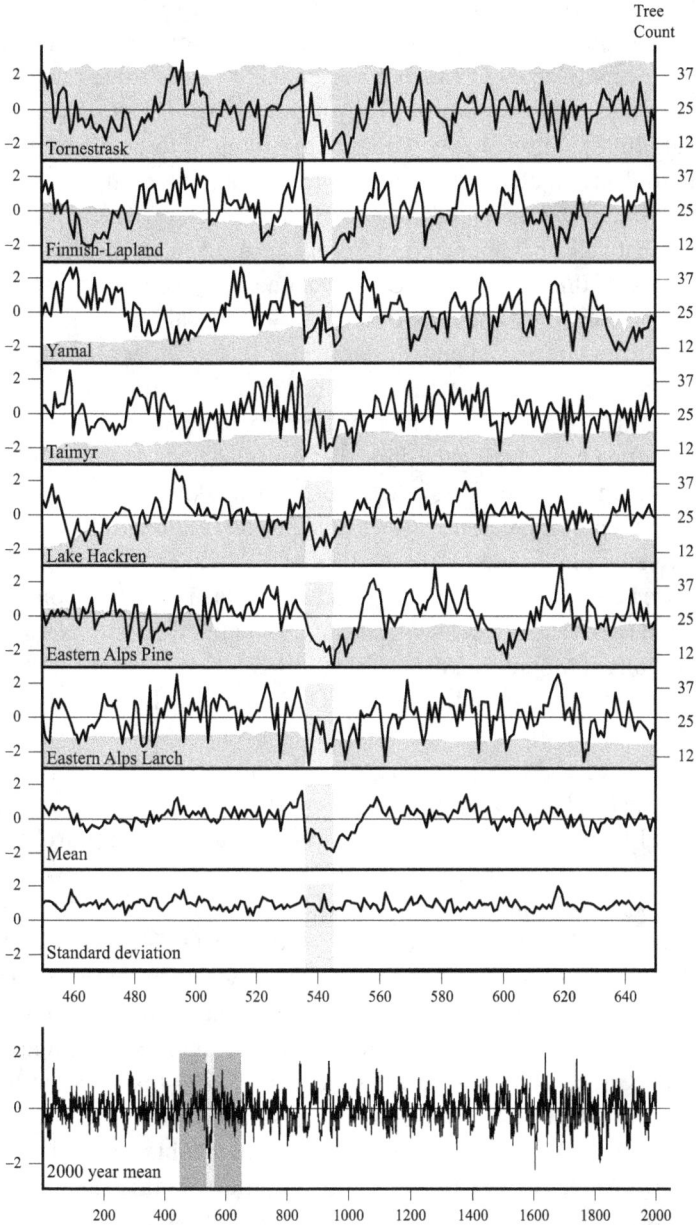

Figure 4.4: Cold period from AD 536 to 545 judging from dendroclimatology

Source: Larsen et al. (2008).

The latter eruption seems to have been larger in scale; traces of volcanic sulfate H_2SO_4 thought to originate from Mount Haruna have been discovered in ice cores collected from three places in Greenland and in Antarctica. They are dated to AD 529 ± 2 (Larsen et al. 2008). The volcanic sulfates are thought to be ejecta from Mount Haruna because of the sedimentation pattern and date, but this kind of detailed analysis has another purpose. This concerns the 'dust veil' that appeared in Europe in AD 536, as mentioned in the title of this section. Based on documentary sources as well as dendrochronology and measurements of chemical impurities in ice cores from Greenland and Antarctica, L.B. Larsen et al. surmised that the black cloud (dust veil) described in sources from AD 536 was caused by volcanic ash from a massive volcanic eruption covering the Earth. On analyzing the data, they revealed that tree-ring growth in Sweden, Finland, Russia and Austria was slow in the years AD 536–545, proving that this was a cold period (Figure 4.4).

Meanwhile, based on the pattern of sulfate sedimentation that showed concentrations of H_2SO_4 in two peaks in ice cores from this period (pattern of dispersal in the atmosphere), they compared the eruptions of Mount Haruna peaking in AD 529 ± 2 with those of the Tambora volcano of Indonesia in AD 533/534 ± 2 (Table 4.1, Figure 4.5, Larsen et al. 2008). Interest in the events of AD 536 started in 1983, when NASA researchers R. Stothers and M. Rampino produced and published a list of all volcanic eruptions in the ancient period from their research on Mediterranean historical archives (1983). This list included the 'continuous dust' that darkened skies for a period of time in AD 536–537. The authors asserted that the reduced temperatures, drought and food shortages caused by the dust were not restricted just to the Mediterranean, but extended over the whole of the northern hemisphere. They continued to write a number of papers, which, however, failed to draw the attention of western classical scholars. From around 1999 this event at last began to attract attention, with the result that many researchers from China to the British Isles and from Arabia to North and Central America have recently started following this catastrophe (Arjava 2005).

On the Korean peninsula, the Three Kingdoms period started from the second half of the third century, but in AD 475 the Baekje capital Hanseong fell to Goguryeo and was moved to Ungjin (present-day Gongju). Then in 538 it was moved further south to

The deposits following the 1809 eruption (unknown source) and the 1815 Tambora eruption

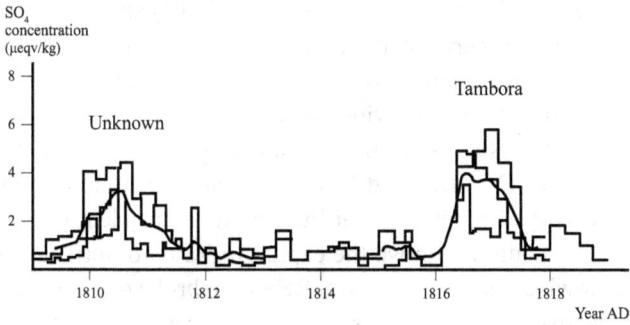

The deposits from what is believed to be the Haruna (Japan) eruption (left) and the eruption that caused the 536 dust veil (right)

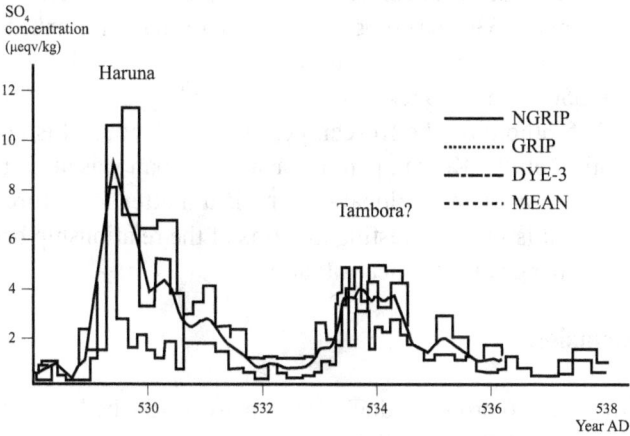

The deposits in the NGRIP ice core following two eruptions of Komaga-Take (Japan) in the seventeenth century

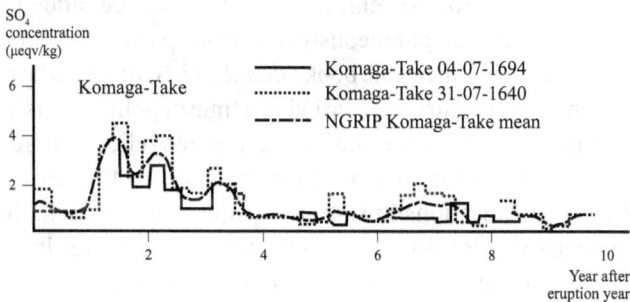

Figure 4.5: Volcanic SO$_4^{-2}$ deposits as registered in the Greenland ice cores

Source: Larsen et al. (2008).

Sabi (Buyeo). The destabilization of the Korean peninsula under pressure from Goguryeo and Shilla had a significant impact on Japan.

In Japan, this period corresponds to the time of the Mount Haruna eruptions mentioned above. This was an era of major political upheaval in the Kofun period. When the Emperor Buretsu died in 506, he left no successor. Instead, a fifth-generation descendant of the Emperor Ojin traveled from Echizen to assume the imperial dignity as the Emperor Keitai. Although he acceded to the throne in 507, he could not move the capital to Yamato until 526. But in the following year, 527, the Iwai Rebellion broke out in Kyushu and there were ongoing problems with the subjugation of powerful local clans. Although there are theories surrounding the confusion after the abdication and death of Keitai in 531, Emperor Kinmei succeeded to the throne in 539 and brought stability to the nation. On the Korean peninsula, however, there were conflicts involving Baekje and Imna, and in 560 (or 562) Baekje was annihilated by Shilla, cutting off Japan's foothold on the Korean peninsula. Whether this political instability on the Korean peninsula and in Japan was at all related to the phenomenon of global cooling is a matter for future study. However, it is very interesting in terms of the relationship between climate change and political situations.

Conclusion

On the AD 536 problem, doubts must arise as to whether a volcanic explosion really could be linked to reduced temperatures on a global scale. One proponent of global history, Geoffrey Parker, has collated various records from seventeenth century Europe, added events that occurred contemporaneously in various parts of the world and published the outcome in his book *Global Crisis*, in which he regards the seventeenth century cold period as a time of political crisis (2013). Parker matches evidence indicating the periodicity of sunspots (Maunder Minimum) and volcanic activity in the seventeenth century with variations in summer temperatures, concluding that volcanic activity is also related to reduced temperatures in summer. In Japan, too, there are similar reports of volcanic activity and climate change. Excavation surveys of the Kamui Tapukopu Shita and Ponma Sites in Date City have revealed a harsh, arid climate with fewer summer days than today in the period between the eruption of Mount Usu (1663) and the Komaga-Take eruption and tsunami

Table 4.1: Selected volcanic H_2SO_4 deposits from well-dated Greenland and Antarctic ice cores

Eruption	Greenland Peak H_2SO_4 Dating[a] AD	Dye-3 H_2SO_4 Dating[b] kg/m²	GRIP H_2SO_4 Dating[b] kg/m²	NGRIP H_2SO_4 Dating[b] kg/m²	Antarctica Peak H_2SO_4 Dating[c] AD	DML05 H_2SO_4 Dating[d] kg/m²	DML07 H_2SO_4 Dating[d] kg/m²
Haruna?	529 ± 2	104 ± 5	101 ± 3	97 ± 3			
536 event	533/534 ± 2	100 ± 7	61 ± 4	57 ± 4	542 ± 17	29.2 ± 6.3	43.9 ± 15.6
1809 event	1810	44 ± 5	29 ± 2	38 ± 3	1809 ± 3	15.4 ± 9.3	27.5 ± 3.4
Tambora	1816	63 ± 4	39 ± 2	46 ± 2	1816 ± 1	32.5 ± 7.0	54.6 ± 5.5

Source: Larsen et al. (2008).

Notes: [a] Greenland ice core dating from Vinther et al. (2006). [b] The natural H_2SO_4 background has been subtracted from all data. The uncertainties in deposited H_2SO_4 are a direct consequence of the uncertainties in the determination of the H_2SO_4 background. [c] Antarctic ice core dating from DI\ fL cores (Traufetter et al. 2004). [d] DML data from Traufetter et al. (2004).

(1640, see Figure 4.5, bottom), while clams unearthed from shell mounds reveal ecological adaptations to a cold environment (Soeda et al. 2012). Volcanic activity seems clearly related to climate change.

There are signs of collaboration in individual research on climate change and disasters on a global scale, as well as the role to be played by diachronic research in HSS, as demonstrated above.

Part II
Human Survivability Studies
Methodology

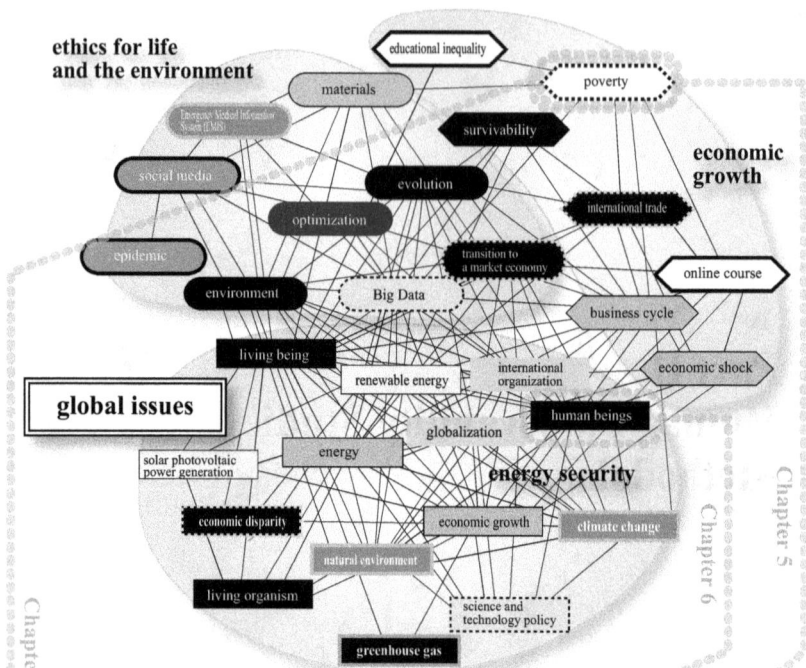

ethics for life
and the environment

educational inequality

poverty

materials

Emergency Medical Information System (EMIS)

survivability

economic
growth

social media

evolution

optimization

international trade

epidemic

transition to
a market economy

online course

environment

Big Data

business cycle

living being

renewable energy

international
organization

economic shock

global issues

human beings

globalization

solar photovoltaic
power generation

energy

energy security

economic disparity

economic growth

climate change

natural environment

living organism

science and
technology policy

greenhouse gas

Chapter 5

Chapter 6

Chapter 7

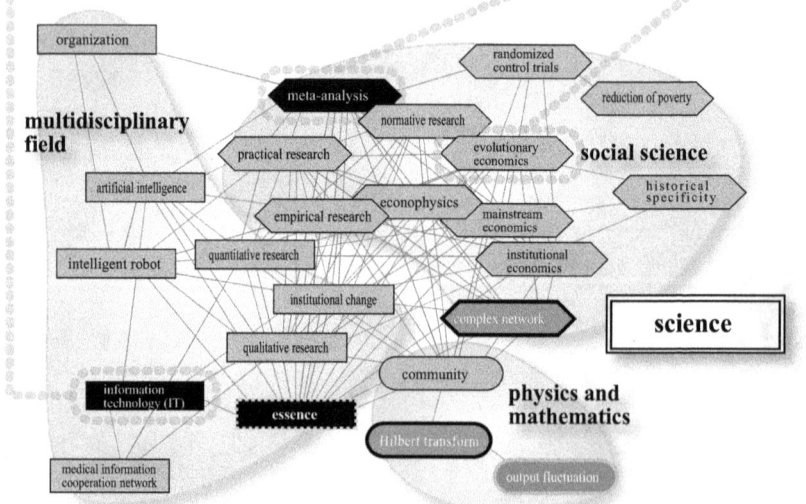

organization

randomized
control trials

meta-analysis

reduction of poverty

multidisciplinary
field

normative research

practical research

evolutionary
economics

social science

artificial intelligence

econophysics

historical
specificity

empirical research

mainstream
economics

intelligent robot

quantitative research

institutional
economics

institutional change

complex network

science

qualitative research

information
technology (IT)

community

physics and
mathematics

essence

Hilbert transform

medical information
cooperation network

output fluctuation

Introduction to Part II

Interdisciplinary dialogue has been initiated toward the establishment of the discipline of Human Survivability Studies. In Part II, we discuss Human Survivability Studies methodology through the topics of economic disparity, innovation, economic growth and energy security from social science, natural science and informatics perspectives.

In Chapter Five, after explaining the significance of policy science that integrates social science research, we offer a new research direction using randomized controlled studies and institutional economics that differ in perspective from mainstream economics. In Chapter Six, we demonstrate evidence of a social law beyond free will based on an econophysics analysis of international trade and energy security. Using meta-analysis as a clue, we present Human Survivability Studies as a kind of collective intelligence. In Chapter Seven, we discuss the possibility of the social implementation of Human Survivability Studies by establishing various information systems, focusing on two aspects: ubiquitous environments and Big Data.

As shown in the network diagram, Human Survivability Studies methodology is structured with multidisciplinary fields: the mathematical and physical sciences are located in the center, surrounded by the social sciences, informatics and environmental science. Further interdisciplinary dialogue is needed for the full establishment of Human Survivability Studies as a new field.

5 From a Social Science Perspective

Hiroshi Sowaki and Dimiter Ialnazov

Introduction

Among the questions addressed by Human Survivability Studies (HSS) are many that natural sciences alone cannot resolve. Dealing with such issues requires the motivation and policy decisions of various entities including international and other organizations, national and local governments, companies and individuals. This process comprises a series of steps: identifying challenges, determining available options and choosing and implementing the selected options.

At each step, natural science knowledge of the natural environment, the structure and nature of materials and technology is required. At the same time, the contribution of social sciences represented by 'empirical research' related to economic, social and cultural issues, as well as 'practical research' related to motivation and policymaking, is essential too. Practical research in social sciences is also necessary to train practitioners with respect to political motivation and policy decisions. Furthermore, there are some examples of value-based, 'normative research' in social sciences such as political philosophy and law that have much in common with humanities. However, to complement the humanities and natural sciences, the social sciences need to collaborate and cooperate with both of them.

Since ancient times, political science that is known for its normative and practical research dimensions has advocated the ideal state of politics. However, empirical research incorporating modern scientific methods can also be used as a basis for political practice. Fishkin's deliberative democracy theory (2009), which is being demonstrated and carried out in countries worldwide, includes normative, empirical and practical research. Economics, which began as the study of governing a nation and providing relief to people, later made use of advanced mathematics to become a science that helps understand and explain economic phenomena. Even though different schools in economics are still at odds, it goes without saying that many of their theoretical findings have been used as the basis for economic policies.

In this chapter, after introducing the social science methodology debate in political science and providing an overview of research utilization theory in policy research, we explain the methodological debate in economics by comparing the research methodology of mainstream economics and evolutionary-institutional economics. Then, we briefly examine randomized controlled trials, a new analytical tool that has become popular in development economics in recent years. Lastly, we discuss what is lacking in conventional social science methods and what HSS has to offer in terms of methodological contributions.

Methodological debates in the social sciences

Pluralistic methods

As the social sciences advanced and became more specialized, empirical research divided into quantitative research, which incorporates mathematics and statistics, and qualitative research, which includes historical research and case studies. Qualitative research began with ethnography and was developed in earnest in the US in the early twentieth century. Since the 1970s, qualitative research techniques such as grounded theory, action research, narrative analysis and content analysis appeared one after another and the use of computers was also incorporated.

The philosophical differences that emerged between positivism, which is the background of quantitative research, and constructivism or interpretivism, which forms the basis of qualitative research, produced heated debates even referred to as 'paradigm wars' (Oakley 2000).

In 1959, social psychologists Campbell and Fiske proposed that multiple tests and techniques be combined to enhance the validity of research (1959: 81–105). Their method, named triangulation, was originally employed in the area of quantitative research. However, since the paradigm debates of the 1990s, mixed methods research transcending the two areas of quantitative and qualitative research has become quite popular.[1] The paradigm behind mixed methods research is pragmatism.

Ikuya Sato (2006: 138–146) made the following observations regarding triangulation: (1) whereas on-the-spot intensive observation is possible in field studies, the number of subjects is limited; (2) in surveys such as questionnaires, involvement with the subject is shallow, small factors tend to be overlooked and the causal relationship is difficult to grasp; (3) although laboratory experiments best clarify causal relationships, it is not clear whether they reflect actual society; and (4) in non-interference techniques, while there is no issue of bias arising from contact between investigator and subject, resources are limited and there is a constraint in terms of understanding the dynamism and detail of actual society. Although no single method is adequate to precisely capture society, it is possible to recognize the advantages and disadvantages of each method and then come up with a suitable combined methodology.

Combined methodology involving natural and social research is also utilized. In the health care sector, in 1998 Greenhalgh et al. proposed the idea that qualitative research is important as a complement to evidence, statistics and scientificity, and now mixed methods research has become common.

High-quality quantitative research clarifies each effect among policy options and enables policy prioritization, while high-quality qualitative research deepens the understanding of challenges and contributes to the process of value formation among policymakers. It is essential to go out into the field when aiming to resolve practical problems. At the same time, however, it is necessary that the information and reasoning obtained from documents, surveys and experiments be used in a comprehensive and transverse manner.

Social science method debates

Empirical research in political science has included quantitative research of public opinion surveys and voting behavior, as well as historical case studies analyzing various political events. History is a one-off event that never repeats, and many individual case studies have researched single events. The landmark research of Allison's *Essence of Decision* (1971), which created three models of policy decision-making, was based on an analysis of American government policy during the Cuban Missile Crisis of October 1962, which only occurred as a one-off event.

In 1994, political scientists King, Keohane and Verba published *Designing Social Inquiry: Scientific Inference in Qualitative Research*, in which they asserted that it is also necessary for case studies to follow the logic of quantitative research in order to be more scientific. This prompted strong objections from qualitative case study researchers, and a debate ensued. Among King et al.'s assertions, the following were particularly contentious: (1) because research by individual observation is not valid to test a hypothesis or theory, the number of observations should be increased; and (2) as the cause of war cannot be ascertained by selecting cases where war broke out, and similarly the cause of the success of democracy cannot be determined by choosing only those countries that succeeded in democratization, such cases ought to be avoided.

In 2004, the arguments put forward by qualitative researchers and King et al.'s counter-arguments were published in Brady and Collier's first edition of *Rethinking Social Inquiry* (2010). Also, 2005 saw the publication of *Case Studies and Theory Development in the Social Sciences*, in which George and Bennett presented a summary of how to proceed with high quality case studies. Brady and Collier raised the following objections: (1) since comparable cases are few in the first place, increasing the number of cases would lead to amassing those unsuitable for comparison; and (2) even in a single case or a small number of cases, a most-likely case or a least-likely case would be beneficial for verifying a theory. However, even King et al. readily acknowledge that the number of cases increases if previous research is taken into account, and that there is significance to triangulation that uses both quantitative and qualitative approaches.

Although King et al. do not necessarily properly appreciate the significance of various qualitative research methods, the debate nevertheless revealed that the process of causal inference is important in qualitative research, and it impacts not only political science but also the social sciences in general.

Policy science and research utilization

The science of decision-making, where research is carried out in relation to risk and organizational behavior from the practical needs of management, is based on fields such as mathematics, information science, sociology, psychology, artificial intelligence, economics and political science.

In response to the specialization of the social sciences and the increased complexity of modern society after World War II, a policy science that comprehensively applied social sciences, such as political science, economics and sociology, to policy was born. Dror (1983) states that if policy science is sufficiently utilized, it is possible to improve the quality of decision-making.

Lasswell, who first proposed policy sciences, defined them as 'disciplines concerned with explaining the policymaking and policy-executing process, and with locating data and providing interpretations which are relevant to the policy problems of a given period' (1951: 14). Policy sciences also include both sciences for analyzing and understanding policies and those for obtaining information that contributes to the determination of policy. Lasswell maintained that policy science consists of both 'knowledge *of*... the decision processes of the public and civic order' and available 'knowledge *in*' those processes (1971: 1).

However, regarding the understanding of the policy process, both Cohen et al.'s garbage can model (1972: 1–25) and Kingdon's policy streams model (1984) show that policy formation does not always proceed in a rational or linear manner. Further, Lindblom and Woodhouse (1968) showed that incrementalism in policy development is also justified by rationalism.

On the other hand, it has long been argued that the results of research in the social sciences, especially sociology, have not been sufficiently utilized in policy (Scott and Shore 1979). Economics, also a part of the social sciences, has been relatively utilized, but this is partly because economic experts have been engaged not only as academics but also as government and business economists. Sociologists, conversely, are almost always limited to academia. However, when systems entrusting policy evaluation to think-tanks are developed, social science research results are more likely to be utilized.

Whether or not the results of social science research actually reach policymakers largely depends on social mechanisms, such as personal exchanges between those conducting the research and those deciding on policy. According to Weiss (1979; Weiss and Bucuvalas 1980: 302–313), whose work is representative of policy science and evaluation, research is seldom used in the 'engineering model' that linearly connects research and policy, but research utilization makes sense in the 'enlightenment model,' which understands that research spreads among policymakers through a variety of routes. In this context the UK government declared 'better use of evidence and research in policymaking' in 1999 (Modernizing Government Secretariat, Cabinet Office 1999: 16) and announced a plan to develop public policy with randomized controlled trials in 2012 (Haynes et al. 2012), funding research in line with these pronouncements.

In science and technology policy, various funding models have been used in each phase, from basic research through to applied and developmental research and practicalization, and even in the steps connecting these phases. However, the humanities and social sciences are not always necessarily targeted. Such policies that bridge social science research and practice are needed, while the unique and enlightening role this research can play should be emphasized in general.

HSS aims at the social implementation of research results, the provision of policy recommendations and even social entrepreneurship. Yet, in order to realize those intentions, a comprehensive strategy based on policy science and research utilization theory is required.

Methodological debates in economics

Research methodology of mainstream economics

As discussed above, one of the main controversies in social sciences is about the use of quantitative vs. qualitative research methods. Mainstream economics (hereafter abbreviated as 'ME') is in the camp of those arguing for the use of quantitative methods. ME uses abstract mathematical models to analyze various economic phenomena. Famous examples include, among others, Solow's model of economic growth and Hecksher-Ohlin's model of international trade. Although the models appear very logical and easy to understand, they have one major disadvantage. Namely, they are based on a number of assumptions such as utility maximization by economic agents and optimal allocation of resources achieved through the market mechanism. Most of these assumptions are far removed from the real world.

Furthermore, ME quite often uses regression analysis. For instance, to estimate which policy should be chosen to achieve the government's objective, the objective is formulated as a dependent variable, and various policy options are shown as explanatory variables (Mitsubishi UFJ Research and Consulting 2014). By running the regression analysis, one could find out exactly which policy has the greatest effect in terms of achieving the objective (Mitsubishi UFJ Research and Consulting 2014). Consequently, the government ought to choose to implement that particular policy and reject the rest of the existing policy options. However, as there are some methodological problems with the use of regressions, the direct application of regression results in the policymaking process is not entirely straightforward.

The use of abstract mathematical models and regression analysis by ME was subject to criticism even before the global financial crisis of 2007–2009, but the scale and frequency of the criticism has rapidly increased since. According to Prof. Paul Krugman, a Nobel Prize winner, the crisis has shown two major failures of ME (Krugman 2012). First, before 2007, the main representatives of the economics profession ignored various symptoms of the impending crisis such as, for instance, the pre-crisis vulnerabilities of the financial system. Secondly, they could not reach a consensus on what the government's response should be after the crisis had occurred. In addition, Prof. Krugman has been a harsh critic of dynamic stochastic general equilibrium (DSGE) models because of their inability to analyze 'low-probability, high-impact' events such as the 2007–2009 crisis.

In spite of their criticism of ME, Prof. Krugman and many others adhere to the prevailing view that if economists could somehow improve their mathematical models and regression tools, they would be better prepared for times of economic disturbance and disorder. In a nutshell, they argue that there is nothing wrong with the methodology of ME, it is just that better models need to be developed to solve more complicated problems. In the following subsection we introduce an economic theory that is radically different from ME, and thus falls into the category of 'heterodox economics'. That theory is 'evolutionary-institutional economics' (hereafter abbreviated as 'EIE').

Research methodology of EIE

The reason we focus on EIE here is that one of the authors (Dimiter Ialnazov) spent many years studying former socialist countries' transitions to a market economy. The transition started in 1989 in Central Eastern European countries like Poland, Hungary and Czechoslovakia, and later spread to South Eastern Europe (Bulgaria and Romania) and the former USSR. During the 1990s, there was a heated academic debate on which economic theory was more suitable for analyzing the post-socialist transformation. Although the debate did not produce any winners, many scholars came to the conclusion that ME was not appropriate for that analysis. Since then, most have been using EIE.

There are at least two main reasons why ME cannot be applied in the study of the market transition. The first is that ME's mathematical models leave 'institutions' out of consideration, or treat them as something exogenous. Institutions are defined as established social rules that can be classified into 'formal' (laws, regulations, contracts) and 'informal' rules (social norms, conventions, mental models). According to a famous definition, institutions are 'the rules of the game in a society' (North 1990: 3). However, any serious investigator of the post-socialist transformation[2] knows it is impossible to leave institutions out of consideration. The transition to a market economy implies a large-scale institutional change that transforms not just the economic system, but also the political, legal, social and cultural systems.

The second reason why ME is considered unfit for that role is that it also ignores genuine uncertainty and processes occurring in a specific historical period (Van de Mortel 2002: 12). We need to distinguish between 'risk' and 'genuine uncertainty'. 'Risk' can be used when we can calculate the probability of a certain event and know the probability distribution. When we cannot do that, we have 'genuine uncertainty'. Under genuine uncertainty, not only can we not estimate the scale of damages or losses as a result of the market transition, but we also cannot know who will be the winners and losers from the large-scale transformation. After the end of the socialist system, individuals, companies and governments had to make decisions and take actions under a situation of genuine uncertainty.

In a nutshell, to analyze the transition to a market economy, we need an economic theory that can take into account institutional change and evolution, genuine uncertainty and processes occurring in a specific historical period. We also believe that compared to ME, EIE is better equipped to explain both the unforeseen consequences of neoliberal market reforms and the variety of national or regional trajectories in the post-socialist world.

In fact, in the early stages of the market transition (1989–1994) when most countries implemented similar neoliberal market reforms, no one could predict the unexpected divergence in the countries' paths to a market economy. However, around the mid-1990s it was evident that the Central Eastern Europeans, the South Eastern Europeans and most of the former USSR (known also as CIS or Commonwealth of Independent States) were headed towards their own distinctive socioeconomic systems (or types of capitalism). To explain these unexpected outcomes, we think that EIE concepts such as 'path dependence', 'lock-in' and 'path shaping' are quite useful.

The market transition experience of former socialist countries has been likened to a social laboratory where we can learn from experiments what exactly is wrong with the ME methodology. What we have established is that the economy should be viewed as a living organism or a complex system. To analyze a complex system, we need holistic approaches that involve systems thinking. This is antithetical to the approach of ME, which is based on methodological individualism and reductionism. Moreover, the solutions proposed by ME seem straightforward and easy to understand, but since they are usually 'one-size-fits-all' solutions, they ignore the distinctive characteristics of countries or regions, as well as the issue of historical specificity. In the end, the solutions proposed by ME just bring about unforeseen consequences and create new problems to solve.

EIE is not only useful for understanding real-world problems like in the case of the market transition discussed above, it also avoids the methodological failures of ME. Yet, few scholars and experts around the world today use EIE's research methodology. In spite of their failures over the years, why have ME and its research methodology remained so overwhelmingly dominant?

According to Hodgson, the reason lies in the magical attraction of the so-called 'general theory' (2001: 4). Similar to the 'theory of everything' in physics, since ancient times social scientists have sought to build a general theory able to explain all social phenomena. In economics, ME continues to dominate the stage because it seems able to play the role of such a general theory. In comparison with ME, EIE can explain why something happened in a given country or region at a particular point in time. However, EIE cannot deliver a general explanation that excludes the consideration of a country's (region's) distinctive context and historical specificity. In addition, as EIE does not use abstract mathematical models, its historical narratives or case studies do not seem very 'scientific'.

Another argument presented by Hodgson relates to the trade-off between the efforts to build a general theory and the analysis of complex systems. According to him, there are only a few examples of general theories that can explain the change and evolution of complex systems (2001: 12–13). Why? Because when one tries to develop a general theory, one has to accept the use of abstraction and simplification. The straightforward and easy to understand solutions proposed by ME are quite popular among policymakers, though in most cases the solutions it presents are just wrong. In contrast, scholars espousing EIE do not accept the use of abstraction and simplification. This explains why the utilization of EIE's concepts and methods remains quite limited in policymaking even today.

Furthermore, some scholars think it may be possible to integrate ME and EIE. The following examples paint a picture in terms of recent efforts in that direction. First, from the side of ME, there have been attempts to include 'institutions' in the modeling. Some scholars have developed an index of institutional quality and have used it as an explanatory variable in their regressions. Second, from the side of EIE, some scholars like Masahiko Aoki (2001) and others have proposed using ME tools such as game theory to analyze the change and evolution of complex systems. This way of thinking has become known as 'comparative institutional analysis'.

On the use of experimental methods in social sciences

In general, it was thought that in the social sciences, in contrast to the natural sciences, it was not appropriate to use experiments to test theories and hypotheses. However, in recent years a new analytical tool called 'randomized controlled trials' (hereafter abbreviated as 'RCTs') has been used quite frequently in development economics. Below we explain how RCTs work and why they have become popular.

The history of RCTs goes back to their use in the medical field to objectively evaluate the effectiveness of new drugs. Patients participating in an experiment are randomly assigned to two groups, a test group and a control group. The new drug is given to the patients in the test group, whereas those in the control group receive a placebo.

Researchers then observe whether the health of the patients in the test group has improved and assess whether the improvement is due to the new drug.

In development economics, RCTs started to be used to evaluate the effectiveness of international development aid projects. Although the goal of development aid is to reduce poverty in developing countries, a large number of empirical studies could not find any evidence that it had really been effective in achieving that goal. The implications were that development aid was a waste of taxpayer money and the donors, i.e. mostly the governments of advanced countries, should reduce or even stop their financial assistance. Particularly since the 2007–2009 crisis, the criticisms coming from within advanced countries have become stronger. This prompted the use of RCTs to identify exactly which development aid projects worked and which ones did not.

RCTs are implemented in a similar way to that described above. For example, in a village in a developing country all women living in poverty are randomly assigned to test and control groups. Those in the test group receive small loans (via microfinance institutions) and basic business training, whereas those in the control group receive nothing. After one year, the researchers observe whether the women in the test group have started their own businesses and as a result, have managed to overcome their poverty. If so, the development aid project (or the so-called 'development intervention') is seen as successful (Keating 2014).

We believe that RCTs may be quite useful for policymakers in general (not only in the area of international development aid) because they can provide evidence about the effectiveness of public policies. Moreover, keeping in mind the methodological debate in the social sciences regarding quantitative vs. qualitative research methods, we can say that by adding more rigor to the qualitative research process, RCTs have the potential to contribute to the integration of natural and social science research methodologies. However, the existence of some robust critique of RCTs should also be acknowledged (see Keating 2014; Council on Foreign Relations 2012). One of the criticisms addresses the problem of the external validity of results obtained from RCTs. In other words, even if results demonstrate the success of a development aid project in one developing country or region, we should not assume that the same project would be successful in a different country or region.

Conclusion

In this chapter, we introduced the social science methodology debate in political science and provided an overview of research utilization theory in policy research. We then explained the methodological debate in economics by comparing the research methodology of mainstream economics (ME) with that of evolutionary-institutional economics (EIE). We also briefly discussed two of the recent efforts to integrate ME's quantitative and EIE's qualitative methodological approach. The results of these efforts, however, have remained very limited. Finally, we examined the recent application of some experimental methods in economics, namely the use of RCTs to assess the effectiveness of international development aid projects.

What is lacking in conventional social sciences methodology? We believe that, most of all, what is missing is dialogue or interaction not just among different disciplines, but also among different schools using different methodologies within the same discipline. Obviously, this is the result of the lengthy obsession with academic specialization that was discussed in the Introduction of this book. Even the recent trend toward interdisciplinary and multi-disciplinary research has proved unable to break down the barriers between fields or within a certain field. In most cases, the meaning of 'interdisciplinary' has entailed simply borrowing the research methods of a different field (school) and applying them in one's own.

Why is dialogue or interaction among different disciplines so important? Because, as stated in the Introduction, the problems that our contemporary society faces have become much more complex and interdependent compared with the past. It has become impossible to find solutions to these problems by using only the narrow specialized knowledge of experts from a certain field. An example of a problem that is cutting across disciplines is the energy transition with its inextricable link to environmental issues. Energy economics and environmental economics that are expected to be interdisciplinary fields capable of offering comprehensive solutions have spectacularly failed at doing that. The main reason is that they have borrowed the research methodology of mainstream economics (ME). As a result, economic growth and efficiency based on narrow cost-benefit analyses have received a higher priority, compared with the preservation of the environment and ecosystems. In our view, the research methodology used by ecological economics that places human wellbeing at the center and regards the economy as a subsystem of the ecosystem is a much better approach to the energy transition.

Furthermore, compared with conventional social sciences methodology, HSS is a new transdisciplinary approach designed to melt the boundaries between different disciplines. It adopts the systems thinking and the holistic perspective that are required to find real solutions to the complex problems of today's world, and yet is also a very practice-oriented academic field. As it aims to link integrated theory with practice, its results should be of practical value to policymakers, managers and social entrepreneurs.

6 From a Natural Science Perspective

Yuichi Ikeda

Why are global issues difficult to contend with?

Vital economic growth is occurring in developing countries that have experienced the economic influence of globalization. On the other hand, a number of global issues have become evident, including the unremitting economic crisis, finite natural resources and energy and climate change. In these global issues, a variety of factors examined by both the social and natural sciences are intertwined in a complex manner. However, solutions to global problems elude us due to the lack of methodological frameworks that integrate the social and natural sciences. In this chapter, I discuss the current state of research on the integration of the social and natural sciences.

Link between natural and social sciences

The natural sciences have developed by dividing objects into relatively simple elements and analyzing these to understand the properties of the objects based on Descartes' theory of reductionism (Descartes 2008 [1637]). Science unveils interesting phenomena and the underlying mechanisms, and consequently our society and lives have undergone dramatic changes that would have been impossible to imagine 400 years ago. Below I attempt to reconstruct our universe using the various elements of knowledge clarified through the sciences.

Basically, as shown in Figure 6.1, our universe has a hierarchical structure. Science on a smaller scale has been conceptualized as providing the foundations of science on a larger scale. For instance, the physics of elementary particles and nuclei, specifically quantum mechanics and thermodynamics, provides the foundations of chemistry, where the properties of organic and inorganic molecules are studied microscopically. Subsequently, chemistry provides the

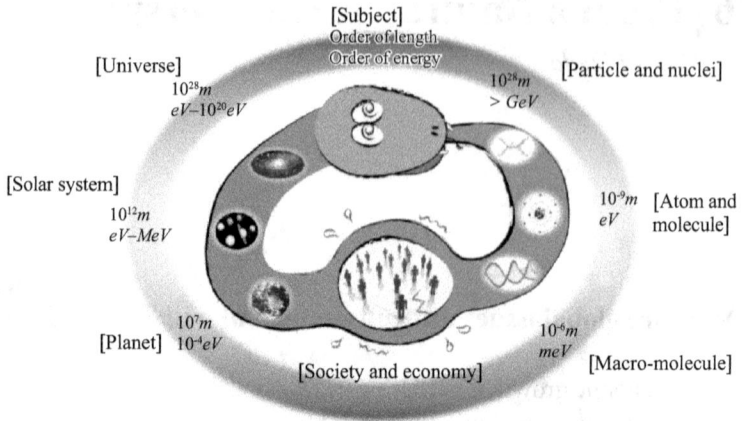

Figure 6.1: Hierarchical structure of the universe

foundations of biology and life science in terms of the study of macro-molecules such as DNA and protein, and their interactions. Next, biology explains aspects of the Earth's natural environment. The foundations of geoscience depend on knowledge of ecosystems and the structural properties of planets. Finally, geoscience explains the planets in solar systems. However, we have singularity here. Which science explains the properties of stars, e.g. the Sun, and of the aggregation of stars, i.e. galaxies? Particle and nuclear physics again plays the crucial role. Knowledge of the physics of elementary particles and nuclei is indispensable for understanding the evolution of the early universe and how stars are born and die. This situation is depicted as a snake eating its own tail, referred to as the Cosmic Uroboros.

Frankly speaking, the abdominal part of the snake that corresponds to society makes me uneasy. The snake has swallowed an egg protected by a hard shell, and is suffering serious digestive problems. I presume that some aspects of society could be the subjects of natural sciences, because society is an aggregation of living matter. However, the science that studies 'society' is entirely categorized under the social sciences today. Consequently, there is almost no research regarding the relationship between the natural and social sciences.

Now, turning to the social sciences, this domain developed very differently to the process undertaken by the natural sciences. It is often said that this difference is mainly due to the fact that precise experiments cannot be carried out using a social science approach. However, similar disciplines exist in the natural sciences, such as meteorology and astronomy, where, like many of the social sciences, only observation is possible. In particular, astronomers have sought to understand their observation of stars by conducting laboratory experiments, e.g. spectroscopy, on matters involving stars, and do not leave their observation results to speak for themselves (Hubble 1936). It is therefore hardly convincing to say that the social sciences are completely different from the natural sciences because of their difficulty in conducting precise experiments.

On the other hand, from the middle of the twentieth century a trend has emerged in academia that seeks to overcome barriers that divide specific academic fields in the process of expanding the scope of research of various sciences. The first signs of this trend were a theory for discussing the formation of living organisms using negative entropy, i.e. the emission of entropy outside a system (Schrödinger 2012 [1944]), game theory to explain the rational and deductive aspects of human decision-making (Neumann 2007 [1944]) and cybernetics to attempt to establish a general theory of control in nature, biological systems and society (Wiener 1965 [1948]). In the aftermath of these pioneering efforts, nonlinear phenomena called 'chaos' characterized by sensitive dependence on initial conditions, strange attractors and self-similarity were found in a variety of hierarchical layers of the universe. The existence of chaos is considered to be the strongest evidence of the breakdown of reductionism. At the same time, the mechanisms of pattern formation due to entropy decrease in a partial system in a state of non-equilibrium were clarified by studies of dissipative structure (Prigogine 1977) and self-organization and synergetics (Haken 1977). Subsequently, the study of non-linear and non-equilibrium systems evolved to complex systems and complex adaptive systems (Holland 1996; Gell-Mann 1995 [1994]), and game theory transitioned to bounded rationality and inductive reasoning (Simon 1996 [1969]; Arthur 1994). Today, a quiet revolution seeking to overcome academic barriers and cross the boundaries of specific fields is progressing through the development of a new strong perspective of complex networks (Watts 1998; Barabasi 2003). However, the current state of play cannot be deemed satisfactory. In the following sections, I explain the germination of Human Survivability Studies (HSS) in terms of the economy and energy.

Econophysics as a prototype of HSS

Establishing ways of overcoming the academic barriers between the natural and social sciences is an orthodox methodology in HSS. Here, we first consider a conventional methodology in the field of economics.

Economics is a social science that can describe economic equilibrium. Economics assumes that economic systems are at equilibrium as a final state after undergoing fluctuations, without knowing whether or not this is the actual case. We know that our economy is constantly fluctuating. Accordingly, it is possible that our economy is far from being in a state of equilibrium, even if it tends toward that state. In addition to this unrealistic assumption, economic theory is basically described by linear equations. However, in reality numerous factors are involved in the economy, and it is necessary to consider the multiplicative relations between these factors. Thus, a basic theory of economy is necessarily non-linear. From this simple consideration, we must note the possibility that the actual economy is non-linear and in a state of non-equilibrium, although economics is a non-linear theory for the state of equilibrium. For this difficulty in economics, we can examine the economy through a different lens. Econophysics is a field of non-linear physics that studies the collective motions inherent in various complex systems using statistical physics, complex networks and agent models. In econophysics, data analysis plays an important role, so it can be positioned as the beginning of data science or big data analysis. More specifically, econophysics is a prototype of HSS.

The economy grows via business cycles. For example, the growth rate of GDP or the index of production shows temporal change repeating positive and negative growth periods over a cycle of several years. This temporal change is a business cycle that occurs due to inventory adjustment. If we take an average of the growth rate of GDP or the index of production for a period involving more than a few cycles, we obtain a positive value. This is the rate of economic growth, and its typical value is a few % for developed countries and 5–10% for emerging countries. We will study business cycles in this section and economic growth in the next.

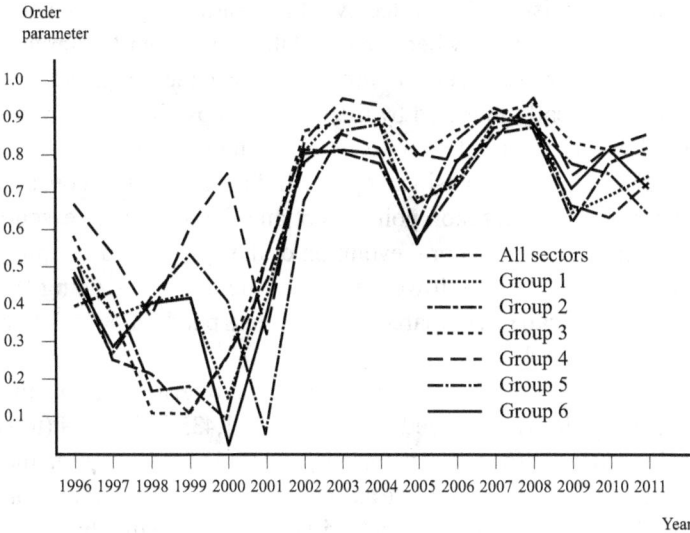

Figure 6.2: Synchronization of the international business cycle

There are various types of business cycles, but the clearest and shortest is that due to inventory adjustment. In the Japanese economy, there are a variety of industrial sectors: the mining, precision machinery, transportation equipment, chemical, electric machinery, construction, wholesale and retail industries. The cycle periods are almost the same, i.e. around four years, for each sector. The question is whether the phase of the business cycle of each industry is coherent or randomly distributed. If we turn our attention to the global economy, we notice that since the early twenty-first century, the industrial structure has changed from a domestically and vertically integrated form to a globally and horizontally distributed one. Consequently, international trade is vitalized toward the regional free-trade zone. How is the relationship between the various industries' business cycle phases affected by this change in circumstances?

Speaking of periodic movement, we recall an oscillator in the natural sciences. The curious collective motion in an oscillator system is known as Huygens' pendulum clock (Huygens 1986 [1673]). Here, two less accurate pendulum clocks hanging on two different columns will show a gradually increasing time difference, but hanging on the same joist, the two clocks will show the same time in perpetuity. This collective motion is due to the interaction through the joist, and is called synchronization. Synchronization is a collective motion where an oscillator system with interaction that is stronger than a certain threshold has a coherent phase, and it is mathematically modeled by the Kuramoto oscillator (Kuramoto 2003 [1984]). It is not easy to imagine the connection between pendulum clocks and business cycles. In fact, it might be confusing to learn that the common point is not an allegory but an essential mechanism of the temporal evolution of the system. How common is an economy in which well-educated people work using their intellects that can be compared to a mind-less pendulum clock? This is a question of degree.

I have studied the synchronization in the international business cycle using the value-added time series of 1,435 industries (thirty-five industries in forty-one countries) and trade dates between these for the last sixteen years. If we assume that the business cycle phase is incoherent, an economic shock will propagate rapidly through the industries. We know empirically that an economic shock that occurs in a certain country is immediately propagated to the rest of the world. For instance, the global economic crisis initiated by subprime mortgage defaults in 2007 and the subsequent bankruptcy of a major investment bank affected by the devaluation of mortgage-backed securities and collateralized debt obligations in 2008 is still fresh in our minds. Does this empirical fact suggest the synchronization of the international business cycle?

The industrial sectors of countries with a large volume of trade are identified through the complex network analysis of international trade. Three stable groups that experienced a high volume of trade between 1996 and 2011 are presented in Figure 6.2. I extracted a phase time series from the Hilbert transform of value-added time series of each industrial sector, and calculated the temporal change of the order parameter of the phase time series. The order parameter is an index that shows the degree of synchronization varying between zero and one, with a larger value indicating a higher degree of synchronization. Figure 6.2 shows that the degree of synchronization is considerably low. Until 1999, the order parameter for each group was about 0.3, which means sectors in each group were weakly synchronized, but there was almost no synchronization across the sectors. In the 2000s, the order parameters increased up to the range of 0.7 to 0.8 for each group in all industrial sectors, although the order parameters in 2005 and 2009 were relatively low. This is evidence of the synchronization of international trade. We know that the amount of international trade has increased monotonically since the 1990s. The volume of trade exceeded a certain threshold in the early 2000s, and consequently the synchronization of the international business cycle emerged as an example of a social law beyond the free will of economic agents. Thus, to a certain degree the business cycle behaves like a pendulum clock, and if an economic shock were to occur in a particular country, its instantaneous propagation might be unavoidable.

Energy, entropy and economy

Economic growth is strongly correlated to energy consumption. Recently, the large-scale integration of renewable energy has been in progress mainly in advanced countries as a last resort for achieving greenhouse gas reduction. In this section, we discuss the effect of the large-scale integration of renewable energy on economic growth.

Electricity is perishable and difficult to store for long periods of time, and therefore supply and demand need to be balanced at each moment. Basically, electricity demand is determined by the total amount of production in the economy. We therefore need facility investment planning and operation planning to satisfy electricity demand for the healthy growth of the economy. Once the supply/demand balance is lost, the frequency of AC power deviates from the standard value. This frequency deviation not only affects the operation of production facilities on the demand side, but can result in nationwide blackouts.

The outputs of renewable energy generation such as solar photovoltaic generation and wind power generation fluctuate depending on changes in the weather in addition to daily and seasonal changes. For this reason, utility companies are not capable of controlling output from their load-dispatching centers. Controlling the output of thermal power plants to absorb the output fluctuations of renewable energy in order to achieve a supply/demand balance is called 'grid integration'. Today, a variety of energy policies have been designed to increase the share of renewable energy integrated into the power grid. However, the increase will be a major factor in breaking the supply/demand balance and will decrease the share of thermal power plants, which play a key role in the supply/demand adjustment in today's power grid. The rise of renewable energy will

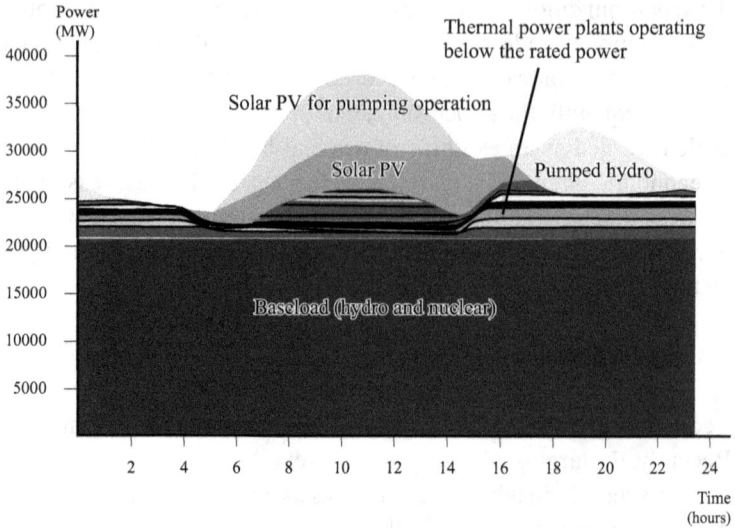

Figure 6.3: Supply/demand balance for the large-scale integration of renewable energy

consequently be a major factor in decreasing the balancing power of thermal power in terms of the supply/demand adjustment. In fact, a variety of issues have been raised in Europe regarding the integration of wind power generation on a scale of several MW.

Figure 6.3 shows the estimated daily supply/demand balance in the Kanto area (Tokyo and the six neighboring prefectures) in 2020 by considering the fluctuation of renewable energies. The portion of the stripe pattern above the baseload depicts the fact that many thermal power plants are operating far below their rated power in preparation for a rapid decrease in output from solar photovoltaic generation. When the share of solar photovoltaic generation increases to about 20%, wide-area operation occurs by enhancing transmission lines and curtailing the excess output from this source. Furthermore, if we want to integrate more than a 20% share of renewable energy, institutional design that maintains a sufficient level at the thermal power plant to balance the supply/ demand adjustment will be required, in addition to demand response equipment and energy storage devices. However, it is concerning that the lack of capacity of the thermal power plants will gradually become evident, because no company invests in facilities that operate once a year, even as existing thermal plants grow older.

Let's consider the relationship between renewable energy such as solar and wind power in terms of entropy. Economic activity is undertaken by producing goods with the input of energy and resources (materials) and by trading goods that are transported to areas of consumption, and finally by emitting industrial waste, air pollution and greenhouse gas. During the transportation of goods, greenhouse gas is emitted. In this manner, the economy is an open system and is in a state of non-equilibrium as a result. The economic system will decrease the entropy within the system by placing materials with small entropy and energy into the system and materials with large entropy outside the system in the same way as living matter. Releasing materials with large entropy outside of the system corresponds to the destruction of the environment, and the accumulative decrease of entropy inside the system corresponds to wealth accumulation, i.e. economic growth.

In general, it is well known that the temporal development of a non-equilibrium system proceeds in order to maximize the entropy production rate, and this is called the principle of maximum entropy production (Sawada 1981). Here, let's assume that we can apply the principle to the economy and discuss economic growth. For the sake of simplicity, we restrict input energy to only two kinds: conventional fossil fuels such as coal, oil and natural gas, and renewable energy as described above.

Solar insolation as an example of renewable energy is widely distributed as an energy density of as low as 1 kW/m^2 on the surface of the ground. This means that the entropy of solar insolation is larger than that of fossil fuel. Accordingly, the usage of fossil fuel is selected based on the principle of maximum entropy production, and the emission of entropy due to economic activity to the outside of the system is maximized.

However, the emission of entropy outside the system results in a decrease in economic activity, introducing renewable energy by policy instead of fossil fuel. Consequently, economic growth slows down by necessity due to the decrease of the entropy reduction inside the system. For this reason, the upper limit of the share of renewable energy is limited by social laws beyond free will, if we set economic growth as a policy target. This is similar to the case of the synchronization of the business cycle. However, we must point out that the integration of renewable energy increases energy security. It is up to each country to choose economic growth or energy security.

Meta-analysis to urgently establish an orthodox method

It takes a long time to establish an orthodox method to overcome the academic barriers between the natural and social sciences. Many believe that global issues are pressing concerns, and that we do not have time to wait for the establishment of convention. Here, I propose an alternative method whereby global issues are tackled by focusing on best practices published in academic papers and reports from international organizations instead of clarifying the relationship between the natural and social sciences theoretically. I assume here that it is possible to extract a methodology (or point towards it at least) to integrate the natural and social sciences by conducting a meta-analysis using academic papers and reports from international organizations. The meta-analysis used in this chapter is a method of text mining to extract the relationships between keywords related to global issues and academic concepts using complex network analysis. I refer to the keywords as 'nodes' in accordance with the usage in network science.

In this section, I outline a case for analyzing research reports written about citizens' understandings of and responses to climate change. This is a typical global issue in which the natural and social sciences are intertwined. A research report consists of a large number of sentences. I extract the nouns as the keywords in each sentence in the report and link them. The attributes of the keywords are provided by a dictionary prepared in advance. I then complete a network consisting of academic concepts as nodes by contracting the nodes without mentioning the attributes in the dictionary.

This academic concept network is a complex network for which the probability distribution of degree (the number of links from each node) is in accordance with the power-law distribution, located in the abdomen of the snake in Figure 6.1. The links from each node are not connected uniformly to the rest of the nodes, but to a specific group of nodes. As a result, the network forms several groups of nodes densely interlinked inside the group, which is called a 'community'. The academic concept network is shown in Figure 6.4 by coloring the communities in green, red and blue in descending order of the number of nodes in each community. The

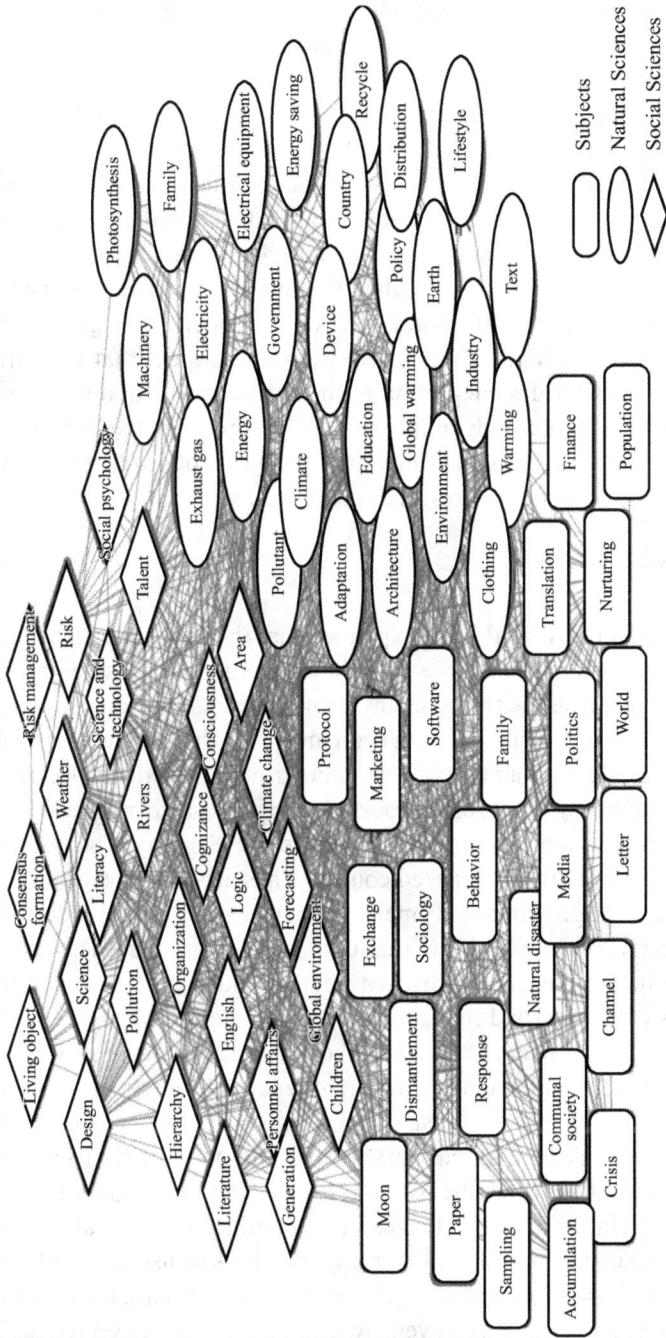

Figure 6.4: Identification of community structure to show the relationship between the social and natural sciences

blue community mainly includes the subjects of academic study. The red community primarily includes academic concepts within the natural sciences and engineering, such as aquatic science, earth and planetary science, environmental studies, resources system engineering and architecture. Finally, the green community includes academic concepts within the social sciences and health science, such as psychology, science education, educational engineering, human information science, nursing and social medicine.

In this manner, it was clarified that natural and social science concepts are used in the study of citizens' understandings of and responses to climate change. We therefore expect that we will obtain hints for the integration of the social and natural sciences through a detailed study of the relationships between the keywords of academic concepts and those of the subjects of academic study using a large amount of research publications.

Current perspective

A recent controversial book explained that the economic disparity between those with a large amount of capital and those without capital increases as the economy grows, because the growth rate of wages is almost on par with economic growth, and the return of capital is higher than that of economic growth (Piketty 2014). The most interesting aspect of the book was the proposal to progressively tax capital as a measure to reduce economic disparity. The rate of economic growth of advanced countries has already reached a low level, and the growth rate of emerging countries will decrease during their economic growth. If progressive taxation is introduced to target the reduction of economic disparity, competition between economic agents will be lost and economic growth will slow as a consequence.

On the other hand, I explained above that economic growth will decelerate through the introduction of renewable energy to replace fossil fuel due to the decrease in the emission of entropy outside the system. From a global perspective, the point of controversy is the extent to which measures for the reduction of greenhouse gases are introduced in advanced or emerging countries at which stage of economic growth. Emerging countries refuse to introduce measures for the reduction of greenhouse gases because it will slow their economies down. Conversely, emerging countries want to use the latest technology introduced as a result of investments or aid from advanced countries to achieve greater economic growth with relatively small levels of greenhouse gas emissions. It has been pointed out that their attitude is logically inconsistent.

Technological and social innovations may continue far into the future, but there are no more economic frontiers due to the development of a global economy. In the future, if the economic growth of all countries converges to a low level, consequently, after sufficient economic growth, the global economy may reach a state of equilibrium, i.e. the fate of economic growth. It is an act of extreme curiosity to ask whether or not we move toward the fate of economic growth and how to reach a state of equilibrium. This is somewhat analogous to the fate of the universe in cosmology. As explained above, our society cannot be determined entirely by our free will, and part of it is determined by social laws beyond our free will. Social sciences such as economics magnify the former aspect, and natural sciences such as econophysics highlight the latter. In the study of human survivability, we need to establish a methodology for integrating the social and natural sciences and conduct practical research aimed at finding solutions to global issues.

7 From an Informatics Perspective

Liang Zhao

Failure to think long-term leads to grief in the short-term

No one knew exactly what had happened to the people in the disaster because there was no way to contact them. At that time, there was no function or system to deal with a disaster or accident twenty-four hours per day in the Office of the Prime Minister. There was no expertise, as the United States has, of computer-aided simulation to respond to earthquakes using the distribution of population, geography, industry and others (Former Prime Minister Murayama in 2006, Wikipedia n.d.-b). TV and radio were the biggest information sources for all the government agencies from the Office of the Prime Minister to local agencies, including other disaster-prevention-related agencies.

It was observed that the reason the Office of the Prime Minister could not get enough information was that the National Land Agency did not develop a method of its own to collect the information, and it failed to collect enough information from other related agencies. It was also pointed out that the information about the disaster area was not fully shared and understood by the whole community – inside and outside the disaster area – which was the reason why the collection and announcement of information was not sufficient after the disaster (Cabinet Office 2007).

On January 17, 1995, a large earthquake occurred in Japan, later called the Great Hansin-Awaji Earthquake. Roads were destroyed and buildings and houses collapsed. A total of 6,434 people were killed, three were missing and 43,792 were injured. The economic damage was estimated at ten trillion Japanese yen (Wikipedia n.d.-b; Cabinet Office 2007).

Failure to think in the long term leads to grief in the short term, as stated above. The Prime Minister at the time, Tomiichi Murayama, later spoke of his regret in a newspaper interview

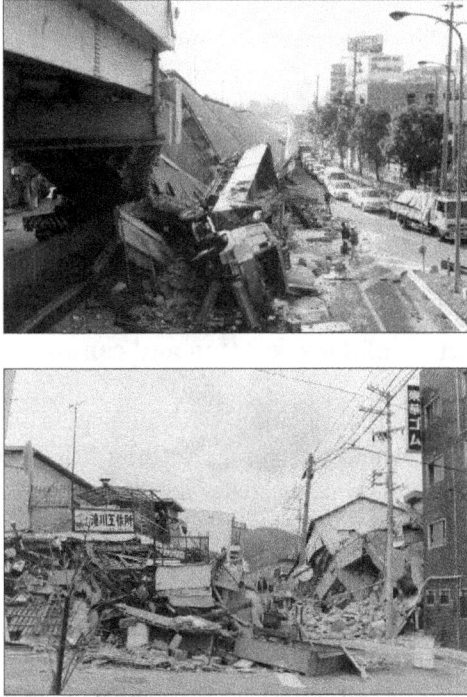

Photo 7.1: The Great Hansin-Awaji Earthquake
Source: http://kobe117shinsai.jp/

(cited above). The initial response was late because there was no information system available for dealing with a disaster of such magnitude. The same problem was also pointed out in 'Lessons from the Great Hansin-Awaji Earthquake' published by the Cabinet Office (2007).

The clear conclusion is that to enable a quick response to disaster, a system for information collection and sharing must first be constructed. Moreover, the government, law enforcement agencies, civil defense forces, medical facilities, mobile service providers, mass media, transportation services and others must coordinate their disaster response efforts. Building such a system requires contribution from all.

To solve large-scale and complex problems, various expertise and techniques must be brought together to interactively seek solutions. This is the goal of Human Survivability Studies (HSS), and its importance has been demonstrated by the response to the Great Hansin-Awaji Earthquake. To solve such problems, we must face these complex issues, motivate HSS and bring global leaders on board. This chapter considers HSS from the perspective of informatics. I provide an overview of informatics using two concepts: 'ubiquitous environment' and 'Big Data'. I then discuss how informatics can help us to motivate HSS, its limitations and future challenges.

Introduction to informatics: Can and cannot

Recently, there have been significant advancements in information and communication technology (ICT). The 'Information Revolution' is often called the third revolution, following the Agrarian and Industrial Revolutions. Various information systems are fundamentally changing our way of life. The environment in which humans exist has already changed significantly. Print publishers and broadcasting services are on the wane. Other traditional businesses such as services, finance, education, medical services, transportation, agriculture and manufacturing all use ICT. Instead of listening to the radio, reading newspapers and watching TV, people are spending time on mobile phones and the Internet. On the other hand, however, people who are not familiar with ICT cannot utilize it and suffer increasing inconvenience.

The origin of informatics can be traced back to Karl Steinbuch, who created the German word *informatik* in 1957 (Widrow et al. 2005). The French word *informatique* was introduced in 1962, the Russian *infromatika* appeared in 1966 and the English *informatics* appeared in 1967. Now it is used to cover the fields that use computers or similar devices, including non-digital sign language and gestures. It has become an integrated field covering the creation, structure, meaning, value and processing of information[1].

The aim of informatics can be summarized as follows: to contribute toward a peaceful and prosperous world via (1) an easy-to-use system for communicating with anybody from anywhere at any time and (2) information and information processing methods that can solve any kind of problem efficiently and correctly. The first, concrete environment is often called the 'ubiquitous environment' and the latter, intangible environment is often associated with the concept

Information revolution

Concrete — Ubiquitous — Environment — Big Data — Algorithms — Intangible

Pyramid (top to bottom): Humanity / Group / Individual / Service / Software / Hardware

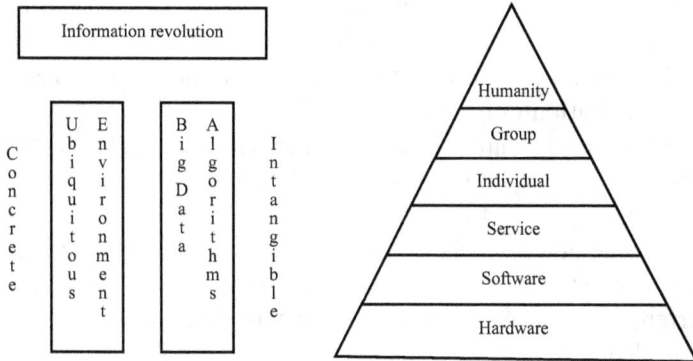

Figure 7.1: Elements of the Information Revolution (left) and the informatics system (right)

of 'Big Data' with the aim of creating significant advancements for humanity (see Figure 7.1, left).

The word 'ubiquitous' refers to an environment in which anybody can connect to anyone from anywhere at any time. It consists of the collection, sharing and exchange of information. Most importantly, the environment can be used without knowing much about the system. In other words, everybody can use it.

Let us take mobile phones as an example. The first 'handy' phone was the TZ-802, introduced by NTT (Nippon Telegraph and Telephone Corporation) in 1987. It was much lighter (900g) than pre-existing products, such as the so-called 'shoulder phone,' which weighed three kg, but it was still too heavy for daily use and never became popular. These days we have 100g mobile phones that are truly 'handy' and have completely changed our lifestyle. In addition, the costs of satellite communication are decreasing more than ever before. As a result, we expect that the dream of a truly ubiquitous environment will come true in the near future.

The informatics system is depicted on the right of Figure 7.1. Under the abstract top layer for humanity, there are several layers: group (from a group of people to an organization or a country), individual, service (e.g., www, SNS, email or online shopping), software (operating systems or various applications) and hardware (e.g., mobile phones, computers and communication devices). Usually, unless an expert in the field, users do not need to know about anything under the service layer to use a system, which is the aim of ubiquitous computing.

However, a ubiquitous environment cannot be realized by communication only. We also need to handle the content data efficiently. For example, the amount of data in Japan increased by 8.7 times in eight years (Soumou-sho 2017) and is expected to increase exponentially in the future. We need not mention that, in a disaster, the volume of data explodes within a short period. Dealing with a huge amount of data in a short timeframe is impossible using the current methods. We need more efficient methods, usually referred to as algorithms for Big Data.

Big Data is a rather new concept first defined by Gartner (2001) as the '3V model', that is, high 'Volume', high 'Velocity' and high 'Variety'. However, the scale necessary to call data 'big' is not clear, and differs from application to application. Among algorithm researchers, some say that it must be at least two orders (that is, 100 times) larger than the current solvable size. By Moore's Law[2], the ability of hardware doubles every twenty-four months, so two orders provide about thirteen years of hardware evolution. Therefore, roughly speaking, the research on Big Data aims to achieve an evolution in three to five years that would otherwise require thirteen years if we waited for the hardware to evolve.

In the 2013 *ICT White Book* published by Japan's Ministry of Internal Affairs and Communications, Big Data, in the strict sense, can be defined as structured data (e.g., customer or sales data) or non-structured data (e.g., audio, radio, TV, newspapers, books, blogs/SNSs, videos/movies, electronic books, GPS and sensor data). In the broad sense, it also includes human resources, organizations and processing methods. It is believed that the research on Big Data will change the entire structure of economics and industry, and hence our way of life. In fact, change has already been confirmed in areas such as manufacturing, agriculture, services, finance and transportation (Soumou-sho 2017). It is also used in other fields, as we will see later.

So far, we have discussed two fundamental concepts of the Information Revolution: ubiquitous environment and Big Data. These concepts have been widely studied in Japan and other countries (Soumou-sho 2017; Wikipedia n.d.-a; Wikipedia n.d.-c). Apple, Google, Facebook, Microsoft, Amazon, IBM and Yahoo are amongst those leading the research in these areas. However, academic research in this regard is also active. For example, Kyoto University, which was the first national university in Japan to open a graduate school of informatics, is promoting informatics-related research in biology, life sciences, cognition science, e-government, multimedia library science, complex systems, the co-existence of human, machine, life and information systems, information communication, network technology, large-scale high-performance information circuits and ULSI (Ultra-Large-Scale Integration), digital signal processing, observations and information processing of the Earth's atmosphere and mathematical and physical methods.

However, we must also be aware of informatics-related problems. For example, current informatics infrastructure, processing methods and services are not satisfactory. The digital divide, cybercrime, discrimination, human rights violations, privacy problems and information control cannot be solved using an informatics approach alone. More fundamentally, assuming that one day computers will achieve superintelligence and become much more intelligent than human beings (I believe that this will happen), what should we do?

There are many examples of the above issues. It is said that increasing numbers of young people cannot live without the Internet. This is good in some sense, but can also be a social concern. Problems related to the Internet occur every day. A convenience store clerk once posted a picture on Facebook of himself sleeping in the ice cream display freezer, and some of his followers copied him in supermarkets and restaurants. In addition, personal information like photos, names, addresses and phone numbers cannot be removed from the Internet once uploaded, and this can exacerbate bullying.

Recently, the *modus operandi* of cybercriminals has shown increasing sophistication and diversification. According to Soumousho (2017), the results of an investigation into the personal damage caused by information security leaks show that the percentage of people 'sure with evidence' and 'possibly sure' that they have suffered damage are 39.4% in South Korea (the highest), 34.6% in France (the second highest) and 24.9% in the US. In Japan, the figure was 15.0% in general and 16.4% when limited to smart-phone users. Another information security issue is that government-led control of the Internet and cyber-attacks may be illegal. In June 2013, former CIA agent and NSA (National Security Agency) contractor Edward Snowden disclosed numerous global surveillance programs run by the NSA, companies and other government agencies[3]. It is also well known that many governments are involved in Internet propaganda, control, filtering and digital wars. These complex issues cannot be addressed by informatics, and other approaches, including politics and legislation, are necessary.

In conclusion, informatics has the potential to support our life and contribute to various fields but, like other technologies, it also has its limitations, divisions and abuses, requiring an integrated approach that can be provided by HSS.

The function and potential of informatics in HSS

We now consider the function of informatics in HSS. ICT can be categorized into the creation, collection, communication and processing of information data. We expect that many ICT applications can solve global issues. For disasters, the advantage is obvious. ICT is also valuable in daily life, for example, in education and to address poverty and inequality issues. If we have the ability to solve large-scale problems that are currently beyond us, we may be able to discover the origin of life, the mechanism of evolution and perhaps extra-terrestrial life as well. We might also travel to and live on other planets. Consider the following examples.

Medical help in disasters

> In the disaster area, whether a person was connected to the Internet or not became the difference between life and death. I felt the weight of the digital divide in that great disaster. (Kan Suzuki, former Senior Vice Minister of Education, Culture, Sports, Science and Technology, in Ohkawara 2012)

It was reported that if a normal level of medical treatment were provided, an additional 500 people would probably have survived the Great Hansin-Awaji Earthquake (Cloud Watch 2015). As a result, Disaster Medical Assistance Teams (DMATs), quick-response medical teams trained to act immediately after a disaster, were established with the aim of saving more lives by preventing avoidable deaths[4].

Furthermore, it was reported that due to inefficient information sharing in the aftermath of the earthquake, the average number of patients per doctor was 3.3 in one hospital, while it was 147.6 in another hospital (Cloud Watch 2015). As a result, the Emergency Medical Information System (EMIS) was created (Cloud Watch 2015; Ohkawara 2012).

A DMAT is a specially trained small team of one doctor, one nurse and one support worker. The purpose of a DMAT is to act quickly in the short-term (usually within forty-eight hours) following a large-scale disaster or an accident involving many injuries. Since these teams were established, they have treated victims of the Niigata-Chuetsu Earthquake (October 2004), the JR Fukuchiyama-line derailment accident (April 2005), the Great East Japan Earthquake (March 2011), the volcanic eruption on Ontake Mountain (September 2014), the recent Kumamoto Earthquake (April 2016) and other disasters.

When a disaster or accident occurs, DMAT headquarters are opened. These headquarters are in charge of overall tasks such as DMAT requests, opening other offices, collecting damage and activity information, arranging for the transportation of injured people, logistics related affairs and planning for the long-distance transportation of the injured if the local medical facility is not sufficient. In the latter case, the DMAT headquarters also take responsibility for patient transport logistics by means such as ambulances, DMAT cars and medical helicopters.

To do this, the DMAT headquarters requires the latest information about hospitals and patients. As well as the Internet and satellite phones, EMIS is used to share this kind of information. EMIS combines the Emergency Medical Systems (EMSs) of each prefecture into one big system with multiple functions including the DMAT management system (added in 2007) and a transportation management system (added in 2010) for patients and aircraft. Based on the hospital and patient information in EMIS, the DMAT headquarters opens other offices and requests or sends DMATs. It also arranges the logistics for and transportation of patients by request from hospitals that have insufficient medical equipment (Cloud Watch 2015).

However, in the Great East Japan Earthquake, mobile phone base stations were destroyed, so neither fixed-line nor mobile phones were functional. Satellite phones also performed poorly. The only available communication method was satellite Internet, but not all DMATs had such equipment, and therefore this was not used in the end. In the worst case, some hospitals with no EMIS functions were forced to struggle on alone. From these experiences, unmanned-aerial-vehicle-based mobile communications (by the National Institute of Information and Communications Technology, Japan), balloon-based mobile communication (by Softbank) and other satellite-based communication methods have been actively researched, and we can expect them to be used in disasters in the future. These new ICTs are valuable not only for medical assistance but also for safety confirmation in disaster situations and volunteer activity coordination during reconstruction. We refer the readers to Katsuyuki Ohkawara (2012) and Hiroki Azuma et al. (2011) for more details.

Solving educational inequality

Art is long, life is short
Learning is considered to be one of the most important human rights. Recently, Massive Open Online Courses (MOOCs) have begun changing the way people learn. A MOOC, like a conventional online course, requires registration. The difference is that it delivers a diploma to those who have completed the courses. These days, many MOOC platforms like Coursera, edX, Udacity, Future Learn, Open2Study and OpenEd have become increasingly popular. Despite the fact that they originated in the West, various educational facilities all over the world have joined in the development (Soumousho 2017; Haber 2014). In Japan, the Japan MOOC (JMOOC[5]) was founded in October 2013. As an example, gacco[6], which was mainly motivated by NTT Docomo, provides many interesting lectures, including one on iPS cells given by the Center for iPS Cell Research and Application (CiRA), Kyoto University, and led by Nobel laureate Shinya Yamanaka. An interesting fact about MOOCs around the world is that many lectures are provided by famous professors, and all are free.

Another new type of ICT-based lecture called the Flipped Classroom is also of interest. In a Flipped Classroom, unlike traditional classrooms, students prepare before they attend by doing tasks like watching videos, and in the classroom, they perform exercises with the help of teachers or other students (Bergmann and Sams 2012). The Flipped Classroom works well with MOOCs. Those who cannot always attend school can take MOOC lectures instead and go to Flipped Classrooms when necessary.

One may wonder how MOOCs can be provided for free. This is possible because the cost of providing an online service such as a MOOC is minimal, aside from the initial cost of creating the video (in contrast, a Flipped Classroom is not free because the teacher's time and effort must be compensated). Thus, everyone can access MOOC courses, provided they have an Internet connection and a device (e.g., a TV, PC, tablet or smartphone) for watching videos. Of course, this assumption, while unproblematic in Japan, may not hold up everywhere in the world. However, given the development of the ubiquitous environment, we can expect it to be true in the near future (free or commercially). MOOCs can also help students obtain higher education at a distance, which should facilitate the alleviation of educational inequality.

Search for extra-terrestrial intelligence

The first day or so we all pointed to our countries. The third or fourth day we were pointing to our continents. By the fifth day, we were aware of only one Earth. (Sultan bin Salman bin Abdul-Aziz Al Saud 2012[7])

By contrast, according to Moore's Law, computers double their speed and memory capacity every 18 months. The risk is that computers develop intelligence and take over. Humans, who are limited by slow biological evolution, cannot compete, and would be superseded. (Steven Hawking 2016)

We have only one Earth, the home where we live. It is inevitable that someday, human beings will need to move from Earth to some other planet. The reason could be the search for new resources or just to escape a global disaster involving artificial intelligence or other events.

For that purpose, we must be familiar with the universe beyond our Earth before we move. Are there any aliens? If so, where are they? Are they friends or enemies? Is there any planet suitable for human beings to live on? If so, where is it? We also clearly need to investigate the origin and evolution of the universe in order to find answers to the questions of the value and destiny of human beings. We still do not know if human civilization is just an accident in the universe or something inevitable[8].

In this regard, there is a series of projects called the Search for Extra-Terrestrial Intelligence (SETI). Recently, in 2015, Stephen Hawking and Russian billionaire Yuri Milner announced new breakthrough initiatives to expand search efforts (AP News 2015). To do this, we need to analyze the radio signals observed by radio telescopes. This task requires a huge amount of computational power. Among many such efforts, SETI@home is a well-known, distributed computing project that searches for the evidence of possible radio signals from extra-terrestrial intelligence using data from the Arecibo Observatory. These data are collected by piggy-backing on other scientific programs using radio telescopes. They are then saved after digitalization and sent to the office of SETI@home (located at the University of California, Berkeley) by post (because the amount of data is so huge it is difficult to transfer it online). There, they are divided into small blocks by time and frequency, then distributed to computers all over the world so that they may search for possible signals that cannot be characterized as noise. The interesting point of SETI@home is that the data, after division into small pieces, are processed by millions of volunteer computers all over the world. The results are then sent to the database, which is maintained by a computer in Berkeley. There, after removing interference signals, the computer tries various pattern-matching algorithms to find the most interesting signals[9]. From this example, we can see that ICT is indispensable for space exploration.

Medical care and drug discovery

> Nine days before Ebola was declared an epidemic, a group of researchers and computer scientists in Boston spotted the hemorrhagic fever beginning to spread in Guinea. By scouring the Internet for clues from social media, local news reports and other available online data, the algorithm developed by HealthMap had an early picture of the deadly disease moving across West Africa. (AFP News 2016)

Early discovery is important for stopping epidemics. However, it is well known that the Japanese government usually adopts a cautious approach when facing a disease or disaster, which makes their announcements somewhat late. This was one of the major reasons for the late initial response to the Great Hansin-Awaji Earthquake.

Fortunately, we now have new tools to aid early discovery. According to the above-mentioned AFP News quote, Big Data is playing an important role in retrieving the signals of disease epidemics. In fact, according to its website, HealthMap, 'a team of researchers, epidemiologists, and software developers at Boston Children's Hospital founded in 2006', is an established global leader in utilizing online informal sources for disease outbreak monitoring and real-time surveillance of emerging public health threats. The freely available website 'healthmap.org' and mobile app 'Outbreaks Near Me' deliver real-time intelligence on a broad range of emerging infectious diseases for a diverse audience, including libraries, local health departments, governments and international travelers[10]. Clearly, with the help of HealthMap and other similar efforts, we will be able to access appropriate information quickly using Big Data and the Internet. We can then announce epidemics earlier and hence arrange for appropriate equipment and human resources to control them at an early stage.

Big Data is also a promising approach in terms of drug discovery. I am currently conducting research on the use of traditional Chinese medicine. We expect that data collection and management, optimization, machine learning, statistics, artificial intelligence and other ICT technologies will help us discover more drugs[11]. Using prescription Big Data, it was observed that almost 60% of medicines can be replaced by generic drugs, estimated to decrease the cost of medicines by about two trillion Japanese yen[12].

Other fields

So far, we have discussed examples of ICT applications. In fact, informatics can be used in every field including science, technology, the arts and humanities. For example, in January 2015, the world's largest museum, the Smithsonian, released into the public domain more than 40,000 digital images of its collections that can be reused for non-commercial purposes[13]. These data are useful for research in the arts and humanities. Big Data is also active in weather forecasting and carbon reduction; for instance, in the recent Green ICT project published by the Japanese Government[14].

In contrast, Big Data can greatly improve our understanding of the human genome, the origin of life and the process of evolution, hence helping us to discover more and more effective drugs, treat difficult diseases and therefore lead a better quality of life. We refer interested readers to (Soumou-sho 2017) for more details.

Future work

'Our Big Bet'

> The lives of people in poor countries will improve faster in the next
> 15 years than at any other time in history. And their lives will improve
> more than anyone else's. (Bill and Melinda Gates 2015)

The Bill and Melinda Gates Foundation is the largest charity
foundation in the world, established by Bill and Melinda Gates in
2000 and later supported by Warren Buffet. The primary aims of the
foundation include four program areas. According to its website, the

> Global Development Division works to help the world's poorest people
> lift themselves out of hunger and poverty. The Global Health Division
> aims to harness advances in science and technology to save lives in
> developing countries. The United States Division works to improve
> U.S. high school and postsecondary education and support vulnerable
> children and families in Washington State. And the Global Policy &
> Advocacy Division seeks to build strategic relationships and promote
> policies that will help advance our work. (Gates Foundation 2017)

In their Annual Letter of 2015, cited above, they made four predictions
about what the world would be like in the next fifteen years with
respect to disease and poverty (Bill and Melinda Gates 2015).
1. Child deaths will go down, and more diseases will be wiped out.
2. Africa will be able to feed itself.
3. Mobile banking will help the poor transform their lives.
4. Better software will revolutionize learning.
The first item is related to medical care and drugs, the second is about
agriculture and the third and fourth are about informatics. The third
item, mobile banking, is a banking system utilizing mobile phones,
smart phones and digital money, whereas the fourth item refers to
the more convenient learning environments, including the MOOCs
and similar innovations that we described above. The Bill and
Melinda Gates Foundation, which has devoted immense effort and
achieved many goals so far in the fight against disease and poverty
in the world, backs their achievements in the following fifteen years
in the above four areas.

There are numerous applications of informatics in other fields; for instance, in communication (automatic translation and interpretation), rescue, exploration, caring with the help of intelligent robots, Big Data and optimization algorithms in decision-making, eyesight or audio support for people with disabilities, digital scanning and imaging in medical care, artificial intelligence in automated driving, digital museums or libraries, discovery of the origin of life and exploration of the universe. We expect that ubiquitous environment and Big Data studies will develop exponentially for some decades to come.

Conclusion

The Information Revolution is changing our world. The construction of HSS is impossible without informatics, which means new possibilities and new challenges as well. To those interested in considering and solving large-scale and complex problems, I suggest mastering the latest ICT environments and Big Data research.

Finally, let's finish this chapter with the last paragraph from the 2015 Annual Letter by Bill and Melinda Gates to all their readers:

The more global citizens there are, and the more active and effective they are, the more progress the world will make. We hope you will show your support by signing up, because we believe that people can and must work together more to make the world a more equitable place. *In fact, we're betting on it.*

Part III
Contemporary Problems and Human Survivability Studies

ethics for life and the environment

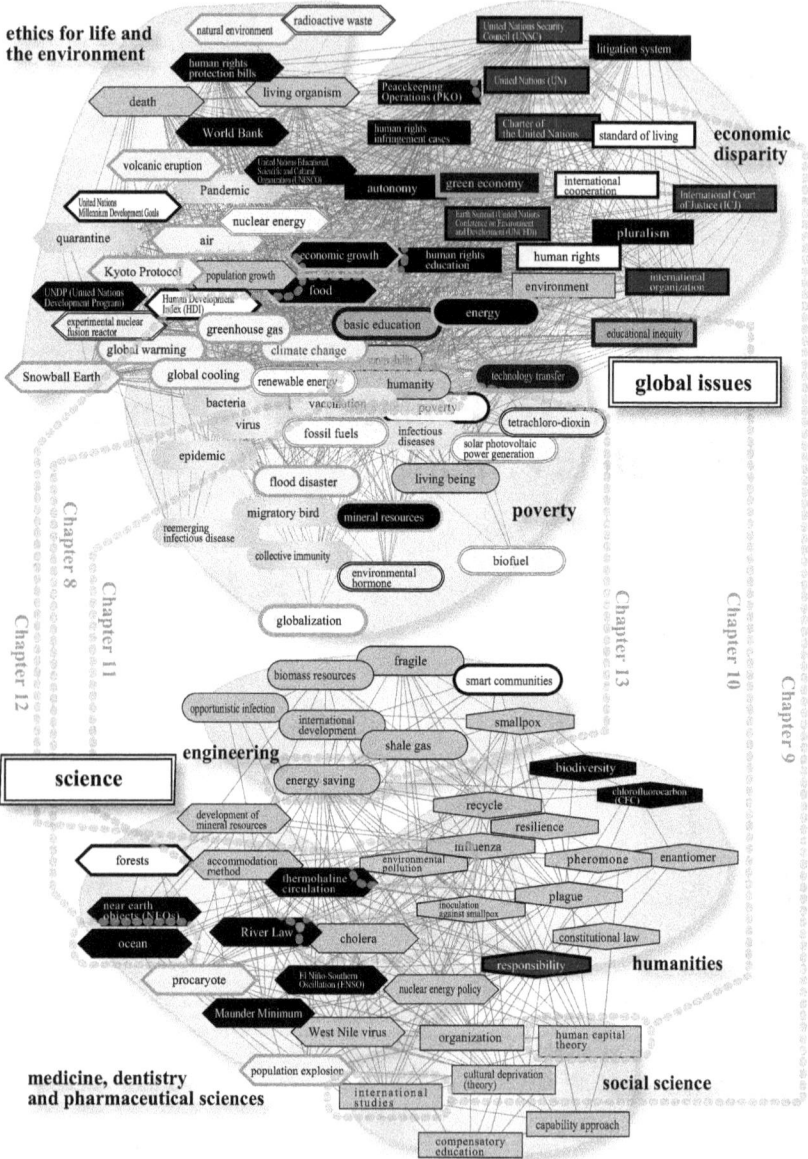

economic disparity

global issues

poverty

Chapter 8
Chapter 11
Chapter 12
Chapter 13
Chapter 10
Chapter 9

engineering

science

humanities

medicine, dentistry and pharmaceutical sciences

social science

natural environment
radioactive waste
United Nations Security Council (UNSC)
litigation system
human rights protection bills
living organism
Peacekeeping Operations (PKO)
United Nations (UN)
death
World Bank
human rights infringement cases
Charter of the United Nations
standard of living
volcanic eruption
United Nations Educational, Scientific and Cultural Organization of NESCO
autonomy
green economy
international cooperation
International Court of Justice (ICJ)
Pandemic
nuclear energy
Earth Summit (United Nations Conference on Environment and Development (UNCHD))
pluralism
United Nations Millennium Development Goals
quarantine
air
economic growth
human rights education
human rights
environment
international organization
Kyoto Protocol
population growth
food
UNDP (United Nations Development Programs)
Human Development Index (HDI)
basic education
energy
educational inequity
experimental nuclear fusion reactor
greenhouse gas
global warming
climate change
Snowball Earth
global cooling
renewable energy
humanity
technology transfer
global issues
bacteria
vaccination
poverty
virus
fossil fuels
infectious diseases
tetrachloro-dioxin
epidemic
solar photovoltaic power generation
flood disaster
living being
migratory bird
mineral resources
poverty
reemerging infectious disease
collective immunity
environmental hormone
biofuel
globalization

fragile
biomass resources
smart communities
opportunistic infection
international development
smallpox
shale gas
biodiversity
energy-saving
recycle
chlorofluorocarbon (CFC)
development of mineral resources
resilience
influenza
forests
accommodation method
environmental pollution
pheromone
enantiomer
thermohaline circulation
plague
near earth objects (NEOs)
inoculation against smallpox
ocean
River Law
cholera
constitutional law
responsibility
humanities
procaryote
El Niño-Southern Oscillation (ENSO)
nuclear energy policy
Maunder Minimum
West Nile virus
organization
human capital theory
population explosion
cultural deprivation (theory)
international studies
capability approach
compensatory education

Introduction to Part III

Part III takes up six key issues in human survival in order to paint a picture of the current situation. Further, we discuss pointers towards solving these issues. Concerning the destruction of the environment, natural disaster and climate change, Chapter Eight examines the causal relationship between environmental destruction on a global scale and natural disaster and climate change. Chapter Nine addresses the issues of ethnic/cultural/religious/state conflict. Chapter Ten explains the current situation surrounding the link between wealth disparity and educational inequality, and considers the cause of this situation.

Chapter Eleven discusses the fight against infectious diseases, predictions of near-future infectious disease outbreaks and risks threatening human survival along with possible countermeasures. Chapter Twelve deals with the population problem, which can at the same time be characterized as the food problem. This chapter explains whether it is possible to produce sufficient food to provide for a future world population of eight to ten billion.

Lastly, in Chapter Thirteen, we focus on resource and energy issues. Here we explore the possible scenario of a shift to renewable energy and examine expectations around the development of a fifth energy source. 'We are all in the same boat' – as we are all passengers on Spaceship Earth, we have to realize coexistence with each other in prosperity. By presenting the current status of the six challenges we face in common, and discussing possible key solutions, we hope this final chapter will form the basis for the discussions in Part V.

8 Environmental Destruction, Disaster and Climate Change

Yosuke Alexandre Yamashiki

Global environmental change and Earth's future

How many years have passed since we reached the point of shouting out that the environmental crisis is on a global scale?

Environmental issues came under serious discussion in the Cold War era, and with the arrival of nuclear warfare, the hypothesized 'nuclear winter' featured. The result of mass destruction by nuclear weapons is the stagnation of aerosols in the atmosphere, decreasing the global temperature by several degrees. The continuation of this process would result in what is called 'nuclear winter', an event that scientists throughout the world have been seriously discussing.

With the collapse of the Soviet Union in 1990, the west experienced a momentary victory. Two years later in 1992, the first and largest UN meeting, the United Nations Conference on Environment & Development (UNCED), was held in Rio de Janeiro, Brazil, concerning the global environment. This international conference was later referred to as the Earth Summit. At this conference, each nation's leaders earnestly discussed global scale environmental problems such as the threat of global warming, the decline of water resources and the decrease in biodiversity. The Earth Summit established two major international frameworks: the United Nations Framework Convention on Climate Change (UNFCCC) and the Convention on Biological Diversity (CBD). In 1997, five years later, a meeting of the Conference of Parties was held in Kyoto that established the Kyoto Protocol. The Kyoto Protocol mandated that Japan, among other nations, reduce CO_2 emissions. Initially, a few problems were identified: the US did not ratify the treaty, there was nothing in place to handle emissions from developing countries and countries as large as China were designated as developing countries. Ten years later, a meeting was held in Johannesburg, South Africa, regarding the possibility of continuing the World Summit on Sustainable Development (WSSD), in which the UN Secretary General Kofi Annan stated, 'After the Earth Summit, progress was slower than expected, therefore, the most important thing is speed. If we miss this chance due to stagnation, there will be nothing but disaster'.

Annan's words lamented the fact that, from 1992, although there was a framework and various discussions around the world, the speed of environmental destruction had increased, while agreements on resolutions advanced slowly.

Another ten years later, in 2012, the United Nations Commission on Sustainable Development (UNCSD) held Rio+20. After the Earth Summit, in order to resolve each country's lack of progress on environmental problems, the discussion focused on economic mechanisms and the introduction of Green economics for global environmental conservation.

However, the G8 summit showed that developed nations are not cooperating in regard to Green economics, and was evaluated as the least successful environmental summit to date. In the meantime, global environmental changes are becoming more and more serious. Especially in recent years, abnormal weather has increased. There is no telling what will happen to the Earth.

According to a recent IPCC report (2014: 6), greenhouse gas emissions have had the largest increase over the last ten years, and forty years ago, the accumulation of CO_2 emissions were about half of today's levels (in other words, CO_2 in the atmosphere has doubled in the past forty years). Also, the written report shows that if the present conditions exceed our efforts, the temperature in 2100 will have risen 3.7~4.8°C (IPCC 2014: 9).

Although the increase over the last ten years is remarkable, the surface temperature of the oceans has largely remained unchanged. This is because the makeup of the surface layer of the ocean is different to what lies beneath it, and the heat is believed to only be stored in the topmost surface layer. This has caused the number of violent typhoons and flooding events to increase. From this point on, despite it being discussed all over the world, humans will experience climate change.

However, aside from the above issues, some scientists say that global cooling poses the greatest danger. They claim the last cold climate on Earth was caused by the Maunder Minimum, a period when solar activity was at its lowest. Today, the Earth may also experience global cooling in a similar way.

Through Miyahara's research that measures the carbon isotopes of tree rings of trees on Yakushima Island (Miyahara 2008: 1380–1382), it has become clear that the occurrence of the lower period of solar activity from 1645–1715 (Maunder Minimum) was part of an eleven year cycle that has since changed to a fourteen year cycle.

Recently, regarding the increasing solar activity cycle, some scientists have argued that this is symptomatic of moving into a new cycle, but at present there doesn't seem to be any sign of it. Simultaneously, in accordance with global warming outbreak expectations, that influence, especially on the ice sheets in Greenland, will cause the north Atlantic to disperse a large amount of cold fresh water, thus the warm climate of Europe would become much colder on the whole (University of Illinois At Urbana-Champaign 2004). Although the salt concentration in the north Atlantic poles is high, the melting of ice sheets will result in the dispersal of a large amount of fresh water, reducing the salt concentration, which will stop thermohaline circulation.

Research on the shutdown of thermohaline circulation and its influence on the twenty-first century was discussed in a recent publication (Vellinga and Wood 2008). According to Vellinga and Wood at the Hadley Centre in England, in the case of the shutdown of thermohaline circulation, the northern hemisphere's average temperature lowers by about 1.7 degrees; especially northern Europe would become very cold[1].

In this way, scientists are maintaining that while global warming is a critical problem, preparation for global cooling must also be carried out. How can we characterize these distinct future predictions?

First of all, it must be recognized that the IPCC AR5 has pointed out that rising greenhouse gases and the global warming mechanism have completely different causes than the above-mentioned cooling mechanism. Global warming is mainly caused by human-made greenhouse gases that shield radiative cooling. Hypothesizing that global cooling is due to solar activity, it may be caused by the declining intensity of solar radiation. However, in previous times of low solar activity, like the Maunder Minimum, the exact mechanism at play and the temperature decline have not been fully understood. Also, since research into the solar activity decline is in its initial phases, a clear scientific mechanism has not yet been established.

Of course, regarding the possibility of thermohaline shutdown due to the melting of Greenland ice sheets, there is a link to the global warming scenario. The IPCC AR4 and AR5 have predicted the strong effects of thermohaline shutdown and more on global warming. However, it is difficult to tell whether a change in the prediction of this effect and fresh water influence estimations are realistic. In Vellinga's research mentioned above, if the estimations are correct, the Earth's equatorial region and polar temperatures will change drastically. Naturally, cooling will occur even if there are other ways for this to happen. One is a volcanic eruption larger than VEI-7. If the ash as volcanic ejecta remained in the stratosphere, the solar intensity on the Earth's surface would decrease. If this continued for a long period of time, cooling becomes a possibility. However, in the first several years, aerosol and sulphuric mist would cause global-scale cooling, but the eruption would also emit a large amount of CO_2, causing global-scale warming for a longer period.

In the absence of these extreme natural phenomena, the IPCC AR5 demonstrates that it will be difficult to escape the effects of global warming caused by human-made greenhouse gases. Therefore, what do humans need to do to 'adapt' to that kind of future?

In this vital matter, humankind has been taking adaptation into account over the past ten years. However, realistically, one could say we are not coping at all.

With the above-mentioned rapid changing of the climate, the global environment will alter considerably. Will humans have the chance to face the challenges this presents?

When it comes to nature, or events arising from the universe, human civilization unfortunately does not have the capacity to manage. That is to say, if there were a temporary change in solar activity, we would have no clue what to do. In contrast, when it comes to anthropogenic environmental changes, will humankind be able to cope?

Humans are known as rational, cognitive creatures, so it is natural for us to think that 'we can restore the changes we make'. However, in reality, with the passing of two world wars, humans with nuclear weapons began to doubt themselves, thinking 'it is impossible to restore the changes caused by our effects'. In the history of the twentieth century, surely the history of 'irreversibility' has been proven. In the IPCC's AR5 (Fifth Assessment Report), various relief scenarios have been proposed to protect against global warming. But surprisingly, these large-scale ideas, more so than in AR3 and AR4, are thought to be mostly impossible to achieve.

Decreasing biodiversity also poses a critical situation. According to the World Wildlife Fund (WWF), since the year 1600, more than 700 species have become extinct. Although the exact number of extinct species is unknown, it is thought that 45~50% of extinct mammals have died out in the twentieth century alone. Also, of all the extinct bird species, 35~40% became extinct in the twentieth century. Though the main cause is overfishing, from now on, these kinds of global changes will increase exponentially. Human beings have already caused significant damage to other living things by destroying the environment.

Regarding the difficulties involved in managing climate change, it is clear that the problem stems from humankind's reliance on burning fossil fuels. Additionally, humans have relied on nuclear power to account for the decreasing dependence on fossil fuels. With the occurrence of two significant nuclear accidents (Chernobyl and Fukushima) and associated radioactive contamination, society is starting to once again doubt whether nuclear energy can provide an alternative to fossil fuels to overcome global environmental degradation.

If humans fail to consider the harm caused by the discharge of CO_2 into the Earth's atmosphere, they will not be able to control climate change. Further, the ability to control radioactive material is not thought possible, and is becoming a significant source of pollution.

We cannot shy away from the environmental disasters and climate change we are facing. We need to try to think about natural disasters and climate change, as well as the effects of environmental destruction and the increased vulnerability of our civilizations.

Examples of catastrophe: Water disasters

Based on the above circumstances, especially using the climate and water disasters of 2010 and beyond as an example, each country is experiencing unexpected torrential rain, and mudslides are occurring more frequently. These disasters are not only occurring in Japan, but have also hit the Philippines and Vietnam, and have caused serious damage especially along the coastlines. In the US, Hurricane Katrina caused New Orleans to crumble in 2004, and the same happened to Galveston in 2008 from Hurricane Ike. In this way, global-scale disasters are increasing, and although they are now labeled 'disasters caused by climate change', other causes were initially identified. Let's look at water disasters as an example.

Japan has a history of floods with a devastating impact on human lives and property. To mitigate these events, Structural Measures have taken place consistently since the end of World War II. Structural Measures are based on the River Law (07/10/1964 #167)[2], and regarding nationally controlled Class A river basins, they protect citizens from floods by managing the responsibility of flood control and building dams and banks where flood risks exist.

Currently, in the River Law: (1) the national government manages the responsibility of flood protection of Class A Rivers, and (2) the national government needs to establish a fixed plan for the purpose of maintaining safety in regards to floods, of implementing Structural Measures and building flood defense systems.

In 1959 there were 5,908 victims (4,697 deceased, 401 missing) of Typhoon Vera (Iwase Typhoon) (Fire Defence White Book 2008). After that, the Ministry of Construction played a central role by investing capital into various technologies on Class A river basins across the country.

As for flood control infrastructure, dams were built upstream that decreased the flow rate, banks were constructed along rivers and estuaries, levee protection works were carried out, estuary dams were built and flood drainage ability was increased by floodway widening and riverbed excavation, in addition to the promotion of cement canal construction, a shortcut for river repair.

New rivers were changed, such as Yodo River in Osaka and Kusatsu River in Shiga. The Kusatsu River was famous for its natural beauty as a Tenjou river (river flowing above the railway and roads); however, the river was replaced by a new channel. Also, reservoirs were established in the circumference of large rivers prone to severe flooding. With this public infrastructure, it is possible to say that Japan's flood risk in terms of Class A Rivers has markedly decreased.

What is the process of establishing flood control infrastructure? Six operational goals are endowed in the comprehensive river plan in Japan: the Flood Protection Plan, Low Water Plan, Environmental Conservation Plan, Erosion Control Plan, Landslide Prevention Plan and Steep Slope Collapse Countermeasure Plan. Based on the most important item, a Designed Flood[3] will be determined based on the

Flood Protection Plan, protecting Japan's Class A river basins with safety measures so they only have to cope with excessive rain once per 100–200 years.

Further, the Forest Act and the Erosion Control Act were developed to defend against landslides in mountain regions. The Department of Agriculture and Forestry played a central role in promoting forest conservation and erosion control, and decreased the dangerous mountain regions in the country.

From this, Japan can meet a period of rapid population and economic growth, and has the basic ability to protect against flood damage in the greater part of the country.

These are the results of thinking about flood disaster mitigation; however, it has reached the point where problems are also being pointed out. First of all, a significant social problem has emerged. Because of dams and dikes, river ecosystems are being destroyed. On the public level, there has been much social opposition to these works starting in the 1990s, and many citizens are opposed to large flood protection infrastructure such as dams and dikes. Conversely, it should be added that there has been a large decrease in Japan's vulnerability to floods.

Also, through flood protection establishment preparedness and the decrease of flood risks, the value of downstream floodplains will increase, and conversely, the cause of large populations gathering in high-risk areas will be identified.

In these circumstances, according to the recent trend in climate change, along with severe weather, the risk of flood events will increase. The future difficulty will be generating acceptable limits. For example, a river dike that would usually experience one torrential rain event per 100 years that causes flooding will, due to global warming, experience this once every twenty years. If this is the case, the river dike's risk of flooding will substantially increase. Furthermore, the flood defense system that accounts for flooding every fifty years needs to plan and mitigate for floods every ten or even five years.

Presently, based on the trend of increasing flood frequency, this kind of example will become more common. In that case, what mitigation measures will we be able come up with? The first thing that can be done is raising the level of cement in man-made river basins. The construction of super levees and reservoirs, as well as the building of multipurpose upper stream dams, are being taken into consideration.

Even though according to citizens, maintaining safety is the thought process behind these initiatives, in reality, there are a number of hurdles to overcome to implement these new measures. The reasons are as follows: (1) the extreme cost, and opposition regarding funding; (2) environmental restrictions – there are soil limitations in the city, making it less feasible to construct cement-sided rivers with such low amounts of soil; and (3) the fact that citizens find it difficult to understand the lack of certainty surrounding possible future disasters. As a recent development, unstructured measures (soft measures) have become a high priority.

These are: hazard map formulation, disaster education in the form of citizen disaster prevention training and using IT to establish a refugee network. Rivers with topsoil banks, however, do not have the power to protect residential areas from actual flooding. Given this fact, how will people escape? This prevention plan doesn't change the fact that flood damage will increase in the future.

In 2014, torrential rain occurred in Hiroshima, causing a landslide and debris flows that resulted in many deaths. This became a problem because a hazard map was not prepared in advance. At the same time, the torrential rainfall occurred at night, the cumulative rainfall was record-breaking, and due to the rainfall build-up, it could not be prepared for completely. Also, torrential rain had never resulted in landslide to that extent before.

Rainfall phenomena are increasing due to global environmental changes, and the Ministry of Land, Infrastructure, Transport and Tourism of Japan is compiling the degree to which this is happening in our country. Firstly, according to statistics from all 1,300 AMeDAS[4] collection points in Japan, there were an average of 209 cases where rainfall events exceeded fifty mm from 1976 to 1985, and an average of 288 cases from 1996 to 2005. This represents an increase of 1.4 times. The average number of cases where the events exceeded 100 mm from 1976 to 1985 was 2.2, and from 1996 to 2005 there were an average of 4.7 cases, a difference of roughly 2.1 times.

Although it's not clear whether the trend would continue if the statistics were available from 2006 to 2015, if one were able to particularly focus on cases of torrential rain in 2013 and 2014, it would appear that the trend is not decreasing.

In the same vein, in São Paulo, Brazil, the number of days in which rainfall exceeded fifty mm in the 1930s was approximately ten, whereas in the 2000s, it was about forty. There were no days in which rainfall exceeded 100 mm in the 1930s, but that has increased to about seven days in the 2000s. The discussion needs to be had regarding whether this is an effect of climate change or the large city 'heat island effect'.

In Japan, in the past fifty years the atmospheric temperature in cities has risen five times more than areas outside the city, and although the heat island effect is a characteristic of big cities, in recent years torrential rain disasters have increased dramatically. The first direct cause is that the Sea Surface Temperature (SST) is increasing in the outskirts. In our country, large typhoons and torrential rain events are caused by rising sea temperatures, and this is related to a La Niña effect. Similar relationships are being discussed as the same type of La Niña and El Niño that occurred in Brazil in 2010 and 2011 that caused catastrophic floods and landslides in Rio de Janeiro's mountain zones.

Recently, it has been reported that, depending on the change in structure of the ocean's water, the El Niño effect (Weng et al. 2007; Ashok et al. 2007; Weng, Behera and Yamagata 2009; Ashok and Yamagita 2009; Weng et al. 2009; Ratnam et al. 2010; Yuan and Yamagata 2014) will create a new form, like the California Niño, which significantly changes the ocean.

Extinctions and disasters

What should we do in Japan to mitigate flood disasters and other risks?

Most people would say that earthquakes and tsunamis come to mind first when they think about disasters. However, compared to earthquakes and tsunamis, floods occur more often, and although periodic, they pose an unpredictable risk.

The question is are enough inspections being carried out? With earthquakes there is risk assessment, and the administration is required to prepare for potential damage caused by floods.

As for Japan, earthquake disasters are recognized as posing a significant risk because (1) they offer meager advance warnings and prediction is difficult, and (2) they cause widespread damage.

On the other hand, earthquakes almost never occur in Brazil, and flood disasters are thought to pose a greater risk. In Japan, you come across weekly articles titled 'The next earthquake will occur in XX', but in Brazil, you more likely to read 'The next flood will occur in XX'.

Risk awareness is seared into the minds of citizens due to incidents such as that of January 2010 where the São Paulo River flooded, completely submerging the small city of São Luis de Paraitinga, whose streets crumbled.

There are three potential causes of the extinction of life on Earth: (A) asteroid attack, (B) gigantic volcanic eruption and (C) climate change. (A) is considered to be the primary cause of the huge extinction event sixty-five million years ago at the end of the Cretaceous period. Other extinction events are thought to be related to this.

It is conceivable that (B) was the principal cause of the Siberian flood basalt at the end of the Permian period. That leaves the last cause of an extinction event, (C). This is thought to be the cause of the 'Snowball Earth', the Cambrian period in which the Earth experienced global cooling. Particularly in modern history, 20,000 years since the last ice age, the Earth's cooling would be a serious natural disaster.

As for gigantic volcanic eruption (B), it had an effect on mankind 74,000 years ago with the eruption of the Toba Caldera in Indonesia. At that time, the volcanic ejecta exceeded 2,800 km³; in Japan the largest eruption was the Aso Caldera, which only produced 600 km³.

In modern history, VEI-8 volcanoes have not erupted; however, various greater volcanic eruptions occurred during the nineteenth century, for example, the 1815 Mount Tambora eruption in Indonesia. It is reported that this eruption even led to crop failures in the US, and is acknowledged as a global disaster. In 1902, Mount Pelee on Martinique of the West Indies erupted, totally destroying the capital San Pierre and killing about 30,000 people. Also, in 1985 the Nevado del Ruiz eruption that occurred in Colombia featured pyroclastic flows that caused mudslides, leading to the almost complete destruction of Almelo at the foot of the mountain, killing 21,000 people.

Although Almelo had prepared a hazard map, this was unsuccessful. The Mayor spread erroneous information about the eruption, resulting in disaster. This highlights the serious nature of the government's judgment during crisis.

In the small country of Montserrat in the Caribbean Sea (British Territory), the capital city Plymouth was abandoned in the mid-1990s after the eruption of Soufrière Hills Volcano because of volcanic ash and pyroclastic flow, and is now in ruins. Volcanic eruptions are rare, but they cause major damage, and should never be taken lightly.

In Japan in 2014, Mount Ontake suddenly erupted, killing sixty mountain climbers. Although there were advance warnings of this eruption, no ban was placed on mountain climbing. Thus, the difficulty of accurately predicting volcanic eruption was again proved.

As for (A), an enormous meteorite collision occurred on June 30, 1908, crashing into the Podkamennaya Tunguska River (Krasnoyarsk Krai, Russia). The meteorite was presumed to be 100 m in diameter. On February 15, 2013, one crashed into the Chelyabinsk River in Satka, with a presumed diameter of seventeen meters. The explosion site was fifteen km away from the city of Chelyabinsk (Popoval and Jenniskens et al. 2013), and caused 5,000 windows to break within the city, injuring 1,500 people.

Although those listed above are the only actual collisions, on June 14, 2002, a seventy-three meter diameter meteorite, 2002MN, came one-third the closeness of the moon to Earth, creating the possibility of a Tunguska-scale explosion. For that reason, Near Earth Object (NEO) monitoring measures were taken at NASA and the International Spaceguard centre, where monitoring for meteors that would affect the Earth continues. In this way, although the possibility of a catastrophic disaster that has the ability to wipe out humankind is low, we cannot forget the risks involved in living on Earth.

Global environmental problems and disasters

Modern civilization directive

In this chapter we accord with the following argument: 'the creation of substances thought to be harmless for the purpose of improving human civilization (CO_2, fluorocarbons, etc.) caused the balance of the Earth to be disrupted, and the possibility of marked environmental changes on the Earth'.

Mass production and mass consumption are the driving force of modern civilization, and in support of the same global values, assets, conveniences and happiness that civilization requires for secure living, stable energy and food security, a large scale natural change is necessary.

In order to create safe living spaces, various river infrastructure and embankments are constructed, and in order to produce stable energy, power plants are built. Over many years, humankind has thought of the Earth as a conquerable object. In modern civilization, the Earth itself is 'dead' and is only an 'environment'. The humans who must live in this world use what is 'usable', 'eliminating' things that are not essential. To these humans who are 'living', they have come to create a 'pleasant environment'.

According to humans, things that we need are pleasant cities and infrastructure, and because of this we also create things we don't need, undesirable things for the Earth, and waste.

This waste includes nutrient salts, organic matter, heavy metals and small amounts of organic pollutants. The atmosphere experiences the greenhouse effect and air pollution; and the soil receives the nutrient salts in manure (N, P), organic solvents, gasoline, small amounts of organic pollutants and heavy metals.

Also of extreme concern is radioactive waste. Although the full meaning of the Fukushima Daiichi nuclear power plant accident is only understood through others' explanations, it is thought that due to forty years of power generation, high level nuclear waste was produced. The meltdown thus resulted in environmental catastrophe.

Of course, although the meltdown and the leak of radioactive waste during the accident in 2011 was unimaginably huge, the Fukushima nuclear disaster can be considered as a consequence of long-term accumulated radioactive waste over forty years' operation.

The Gaia hypothesis on human induced climate change

Climate change problems and other pollution/contamination primarily affect the Earth's ecosystem in a negative way, and greenhouse gases destroy Earth's entire temperature balance. Using human beings as an example, our appropriate body temperature is thirty-seven degrees Celsius, anything above or below that can hinder our organ functions. For that reason, body temperature works as a self-regulating mechanism; in hot weather we can reduce out temperature by sweating, and if it becomes cold we can work to increase it. Assuming this is the case for the Earth, in order to preserve the fixed temperature of the Earth's surface, the ocean current moves the heat to the polar regions, and through evaporation, heat can be released.

In the greenhouse effect, because the Earth's balance is disrupted, warming can't be stopped, leading to various problems.

As a result of the Earth's body temperature increasing (atmospheric temperature), various parts (areas) and internal organs (forests/polar ice caps/ocean currents) are liable to be destroyed. In other words, we should treat global warming not as simply an increase in body temperature, but as if the Earth's body temperature has risen to the extent that its internal organs are collapsing (Earth as a life form is dying), and the environment is changing completely.

Although humans think of the Earth as a fixed ground upon which to amass their civilization, devouring it greedily, humankind is one part of this life form called Earth. This part is abnormally multiplying, causing Earth to be in a critical condition. In other words, so long as we continue in this way, humans are like cancerous cells on Earth's surface. According to James Lovelock, Earth is assumed to be one life form (Gaia). By nature, life possesses a 'self defense mechanism' to avoid extreme conditions. In the current state of affairs, what is this life form's biggest danger? The answer would unmistakably be the existence of humanity.

Because humankind lacks the ability to control the measure of 'civilization', from every single angle, humanity poses a threat to Earth as a life form. Furthermore, humans have rapidly expanded the scope of their lives, causing serious damage. So, what would be the disease control prevention measures Earth would take? Make no mistake, if this was the case 'the eradication of humans' would be the course of action. Thus, if civilization remains as it is now, the Earth as a life form will either be destroyed, or, the human civilization on it will be destroyed.

The contradiction of engineering and Gaia thought

On the contrary, assuming for a moment that we can understand ideas like the above, if we try to think in terms of engineering concepts, the first thing should be 'human prosperity' instead of 'human life'. Even though in order for Gaia to live, a large number of humans would have to die, human engineers are thinking every day about how humans can survive.

In the disaster prevention field, in order to save even one person (the smallest possible goal) you must remodel nature, by for example constructing river dikes. However, the only way humans can survive is through 'nature restructuring', and in this sense we can understand that it is impossible for 'Gaia' to survive as well. This problem has serious inconsistencies that we must cope with. Because of this type of infrastructure, flood defenses like embankments and dams will unmistakably cause injury to Gaia. The same goes for building embankments to protect against tsunamis.

Of course, we may avoid building disaster mitigation infrastructure because of its serious impact on the natural environment, on Gaia itself, or as a result of high estimated cost. However, realistically, without this infrastructure, humans would suffer. When protecting against disasters, we should think about ways we can be less dependent on infrastructure and instead use new technologies, like hazard maps, as 'evacuation models'. We need to keep in mind that the hazard map does not protect us directly, but only shows us how to escape from disaster.

Considering for a moment the prioritization of Gaia's life, this would entail giving up most of the 'blessings of modern civilization'. Would it be possible to do so? How do we think about and take action towards the contradiction between 'prioritizing Gaia's life' and 'our dependency on modern civilization'? Although the solution does not yet exist, we should be searching in this direction.

The survival of humankind

From this point forward humans need to think deeply about considering and respecting 'Gaia, the Earth as a life form'. Especially, before we categorize our survivability problems as environmental problems, energy problems and food problems, we should examine how we can preserve Gaia's stability over the long-term. Since the Earth Summit, there have been other summits and international conferences involving heads of state from each country. However, the concept of 'Gaia Earth as a life form' has not gained much support.

It was made clear at the 2012 UNCSD Rio+20 that implementation of green economics has not occurred due to the economic problems of each country. In other words, more effort was made toward dealing with the problems of modern civilization than examining or enforcing the green economy.

Within the laws and organizations that are human institutions, the concept of 'Earth as a life form' or 'Earth as an individual life' has not been fully clarified. Accordingly, a statesman representing 'Earth as an individual life' to protect her interests does not exist.

The 'Earth as a life form' principle is not yet sufficiently understood, the criticality of the situation has not been recognized and there is no consensus on how to obtain compassion for 'Earth's survival' from all creatures living and depending on her. Accordingly, the 'Earth' continues under the 'conditions of a disease', which will become worse unless drastic action is taken.

For this purpose, we need to construct an academic framework that earnestly thinks about and acts on how humans live on Gaia, and seriously consider Gaia's future as our shared destiny. The new academic framework called Human Survivability Studies can be summarized as the study of 'how we can cohabit with Gaia'.

9 Ethnic, Cultural, Religious and International Conflicts

Makoto Ohishi

Introduction

Conflicts and disputes

The human being is endowed with reason: as Pascal phrased it, a 'thinking reed'. Humans are thus thought to be fundamentally different from other animals. Nevertheless, in the same manner as in the animal world, human society has been shaped by constant conflicts and disputes since ancient times. Starting with disputes between individuals, various conflicts continue to arise on all levels, including individuals, groups, regions, centers and states, etc. Conflicts and disputes among these subjects can be caused or triggered by ethnic[1], cultural[2] or religious[3] differences.

Basic perspective

Ideals of or hope for a world without conflicts and disputes do not solve this situation. On the contrary, this sort of idealism often leads to fundamentalism, where people believe in the existence of an absolute truth and which, therefore, tends to aggravate conflicts even further. Given this state of affairs, we should instead face the fact that we live in a world where there will always be some form of conflict. More specifically, it is rather important to explore the causes of these conflicts and disputes, formulate efficient systems to resolve or settle them and effectively implement said systems.

Next, based on this view, I examine the various conflicts and disputes that surrounds us, and track the numerous efforts to formulate and implement measures to solve them, focusing mainly on the legal and political systems.

International conflicts and disputes

International conflicts

First, at a macro-level, when we take a look at the international community, which consists of 193 countries that are members of the United Nations (UN), we can see a variety of conflicts and disputes, from skirmishes over territorial possession to international wars.

Since October 1945, when the UN was established with fifty-one member states in the wake of World War II, the international community has experienced numerous conflicts and disputes. Some prominent examples include: the 'Cuban Missile Crisis' of October 1962, which was the closest the world came to nuclear war during the Cold War between the Western and Eastern Blocs; the Yom Kippur War of October 1973, fought by Egypt, Syria and other countries against Israel, and part of a series of confrontations that occurred after the state of Israel was formed in 1948; and the Falklands War (*Guerra de las Malvinas*, in Spanish), which began in March 1982 after the invasion of British overseas territories (the Falkland Islands and South Georgia) by Argentina.

Even after the cessation of the Cold War, disputes have continued to occur all over the world. In particular, the Gulf War (1990–1991), which began with Iraq's occupation of Kuwait, and the ongoing long-term confrontations between armed groups from Israel and Palestine, remain fresh in our minds. Even more recently, we cannot forget the conflict between Russia and Ukraine over the annexation of Crimea (2014), and that between so-called Islamic State (ISIS) and Iraq, Syria and Western countries.

Notwithstanding this fact, according to the United Nations Information Centre's explanation, during this period, one of the 'achievements of the United Nations and its organizations' is that 'By sending a total of 64 peacekeeping and observer missions to the world's trouble spots over the past 60 years, the United Nations has been able to restore peace and reconstruct numerous countries after conflicts'[4].

Ethnic and cultural conflicts

In addition to the above conflicts in the international community, European countries that appear at first glance to be stable also have various ethnic and religious domestic conflicts.

For instance, England (The United Kingdom of Great Britain and Northern Ireland) has experienced an independence movement in Scotland, which has maintained its own legal, educational and religious system for centuries, and conflict (including armed conflict) between Catholics and Protestants that formally began with the independence of Northern Ireland from the Irish Free State (Republic) and subsequent annexation to England. The Scottish independence issue was settled, to a certain extent, by the referendum on September 2014, in which voters said 'No' to independence. However, England experienced violent conflicts with terrorist attacks carried out by the Irish Republican Army (IRA) in the 1970s, and Spain has been troubled by terrorist attacks at the hands of separatist group ETA (Basque Homeland and Liberty), which over time increased its use of violence.

Nevertheless, it seems that the situation in the Basque Country is improving. The Spanish Constitution of 1978, established under the Constitutional Monarchy after General Franco's death (1975), not only placed 'political pluralism' as one of the highest values of its legal order (art. 1) and recognized the right to self-government of the nationalities and regions (art. 2), but also allowed provinces with 'common historic, cultural and economic characteristics' to form 'Autonomous Communities' (art. 143) and granted the Chartered Community of Navarre and the Basque Country broad autonomy. With the new constitution, the situation was gradually controlled, and as the high unemployment rate slowly decreased, terrorism was controlled and Spain started to move towards social stability. In 2011, ETA declared a permanent ceasefire.

In England, in terms of the conflict with the IRA, the Good Friday Agreement (Belfast Agreement) between England and Ireland was approved in April 1998 after the British Government adopted a policy to allow the residents to make a democratic decision about which country they wished to belong to. After the Agreement established the Northern Ireland Parliament and North/South Ministerial Council (NSMC), comprised of

ministers of the said parliament and representatives of the
Republic of Ireland, the situation has been improving little
by little.

Conflicts and disputes in Japan

Litigation and the courts

At a micro-level, when we examine the conflicts and disputes in
Japan, we can see that although Japan is not a 'litigious society' to
the same extent as the US, it has a considerable number of legal
disputes. For instance, according to the 'Judicial Statistics' of 2014,
the total number of new civil and administrative cases filed with
summary courts, district courts, high courts and the Supreme Court
(excluding those filed with family courts) totaled 1,455,723. If we add
the total number of cases (910,648 cases) filed with family courts,
which handles the adjudication and conciliation of domestic relations,
we may say that a substantial number of cases are being ruled in
courts across the country[5].

Human rights infringement cases

Besides the cases handled by the courts, there are also a variety of,
so to speak, everyday problems, of which school bullying is a classic
example. This can be easily understood by looking at the report on
'Human Rights Infringement Cases' published by the Ministry of
Justice every spring. This report is a compilation of the achievements
of the human rights organs of the Legal Affairs Bureau and District
Legal Affairs Bureaus, which receive the complaints of persons
who have had their human rights violated and work to relieve the
respective infringement based on the Ordinance on the Investigation
and Resolution of Human Rights Infringement Cases (Item 2 of the
official directive of the Ministry of Justice of 2004).

Note that the Ordinance on the Investigation and Resolution of
Human Rights Infringement Cases is not a law – it is just a directive
of the Ministry of Justice. Therefore, the measures specified
in this directive cannot be imposed on citizens or individuals
by force (and, as I will describe later, this is why the Human
Rights Protection Bill was formulated, giving coercive powers to
certain measures).

If we look at the status of Human Rights Infringement Cases in 2015, by adding the number of new relief procedures filed in that year to the number of pending cases of the previous year, we notice that the total number of cases handled in 2015 was up to 21,044 (94.3%). Furthermore, we should note that one of the characteristics of the new relief procedure cases filed in 2015 was that the number of cases related to (1) human rights infringements on the Internet, (2) labor rights and (3) bullying in schools is increasing, and when compared to the previous year, (1) and (2) have particularly increased by 21.5% and 10.8% respectively[6].

Regarding the use of the term 'human rights', it should be noted that the term as used in human rights infringement cases does not have the same meaning as that used in constitutional law, which refers exclusively to the individual rights and freedoms that should be guaranteed in relation to governmental power (organs of the state, local governments, etc.). If we look at the 'Human Rights Infringement Cases' report of the Ministry of Justice mentioned above, we can see that among the 20,999 new relief procedure cases filed, only 6,043 (28.8%) are cases of human rights infringement 'related to a public servant acting in an official capacity', and the remaining 14,956 cases (71.2%) are related to private law disputes, i.e. disputes between individuals, rather than to public law.

Efforts towards dispute conciliation and resolution

The United Nations Security Council (UNSC)

The most important system to resolve international disputes is that centered on the UN, which was founded on October 24, 1945 after the end of World War II. The Charter of the UN states that the three purposes of the UN are (1) maintaining international peace and security, (2) developing friendly relations among nations and (3) achieving international cooperation in solving international problems of an economic, social, cultural or humanitarian character, and in promoting and encouraging respect for human rights and for fundamental freedoms (art. 1). In order to achieve said purposes, it established six principal organs, namely, the General Assembly, the Security Council, the Economic and Social

Council, the Trusteeship Council, the International Court of Justice and the Secretariat, together with many other subsidiary organs.

Among these, the organ that plays a crucial role in resolving international disputes, i.e. achieving the purpose of maintaining international peace and security, is the Security Council (UNSC). While other organs of the UN are vested only with powers to make recommendations towards member states, only the UNSC is vested with powers to impose obligations on them, because all members of the UN have to agree to accept and carry out the UNSC's decisions (art. 25 of the Charter of the UN).

At present, the UNSC is made up of five permanent members, namely, the US, England, France, Russia and China, and ten non-permanent members. The most important thing here is the voting system, in which excluding procedural matters that are made by an affirmative vote of nine members of the council, all substantial matters are made by an affirmative vote of seven members including the concurring votes of the permanent members (art. 27 of the Charter). This means that a negative vote by a permanent member prevents the adoption of a proposal, even if it has received the required votes; therefore, permanent members' negative votes are called 'vetoes', meaning that permanent members are vested with the power to reject a decision.

In fact, the reality is that all permanent members have cast vetoes depending on the content of the proposals. According to a report by the Global Policy Forum, the veto power has been exercised 269 times from the UN's date of foundation to 2012. For that reason, the UNSC has received criticism for its ineffectiveness in maintaining or restoring international peace.

Nevertheless, the UNSC can demand that the parties to any dispute, the continuance of which is likely to endanger the maintenance of international peace and security, seek a solution by negotiation, enquiry, mediation, conciliation, arbitration or judicial settlement. They can also mandate that the parties resort to regional agencies or arrangements, or other peaceful means of their own choice, and may investigate if the continuance of the dispute or the situation is likely to endanger the maintenance of international peace and security (art. 33 and 34 of the Charter). Furthermore, the Council is equipped with extremely comprehensive powers that include: determining the existence of any threat to peace, breach of the peace, or act of aggression and making recommendations about what measures shall be taken to maintain or restore international peace and security; demanding that member states take measures that include the interruption of economic relations, rail, sea, air, postal, telegraphic, radio and other means of communication, and the severance of diplomatic relations; and, demanding that the air, sea, or land forces of member states take actions that may include demonstrations, blockades and other operations (art. 39 to 42 of the Charter).

Note that sometimes when a permanent member does not support a proposal but does not want to prevent it from being adopted by casting a veto, said member abstains from the vote, allowing the UNSC to win the nine votes necessary for the approval of the proposal.

UN peacekeeping activities

One of the peacekeeping activities of the UN is its Peacekeeping Operations (PKO). The PKO includes a variety of activities (for example making sure that the concerned parties of the dispute are complying with agreements concerning the prevention of the recurrence of armed conflicts, etc.) that are based on UN resolutions and overseen by the UN, which are conducted in order to handle disputes and maintain international peace and security. Because it is not clearly specified in the Charter, it is said that the PKO is in 'Chapter Six and a Half', between the methods of resolving disputes peacefully under Chapter VI and more forceful military methods under Chapter VII.

Under the Cold War between the US and the Soviet Union, peacekeeping by the Security Council failed to function properly. For that reason, peacekeeping activities were carried out mostly by small and medium state members of the UN. As shown in Figure 9.1, even now, most peacekeeping operations are conducted in Africa, the Middle East and Europe.

Note that the agreement of state parties to the dispute and the request of the Secretary-General of the UN are prerequisites for conducting these activities. The most common operations are election monitoring, road reconstruction, etc., but there are also other operations such as cease-fire monitoring and disengagement observation.

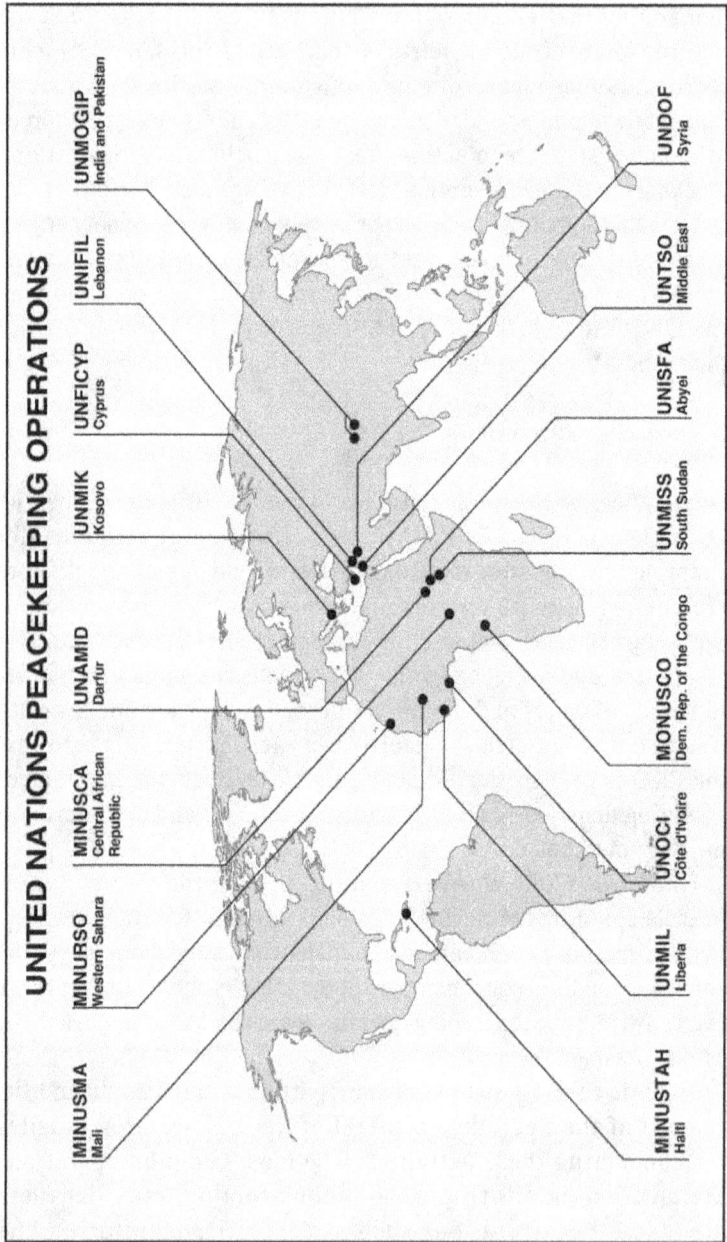

UNITED NATIONS PEACEKEEPING OPERATIONS

MINUSMA
Mali

MINURSO
Western Sahara

MINUSCA
Central African
Republic

UNAMID
Darfur

UNMIK
Kosovo

UNFICYP
Cyprus

UNIFIL
Lebanon

UNMOGIP
India and Pakistan

UNDOF
Syria

UNTSO
Middle East

UNISFA
Abyei

UNMISS
South Sudan

MONUSCO
Dem. Rep. of the Congo

UNOCI
Côte d'Ivoire

UNMIL
Liberia

MINUSTAH
Haiti

Peacekeeping operations since 1948............ 71

Current peacekeeping operations............ 16

PERSONNEL

Uniformed personnel............ 104,773
(89,546 troops, 13,434 police and 1,793 military observers)

Countries contributing uniformed personnel............ 123

International civilian personnel (as of 31 July 2015)............ 5,256

Local civilian personnel (as of 31 July 2015)............ 11,215

UN Volunteers 1,809

Total number of personnel serving in
16 peacekeeping operations............ 123,053

Total number of fatalities in all
UN peace operations since 1948............ 3,466 *

FINANCIAL ASPECTS (US$)

Approved budgets for the period
from 1 July 2015 to 30 June 2016   About 8.27 billion

Outstanding contributions to
peacekeeping (as of 30 June 2015)   About 1.6 billion

*Includes fatalities for all UN peacekeeping operations, as well as political and peacebuilding missions.

Source: http://www.un.org/Depts/Cartographic/map/dpko/P_K_O.pdf (Factsheet: March 31, 2016)

Figure 9.1: United Nations peacekeeping operations

Judicial system and other measures

Litigation system and the right to the courts

Next, I examine the situation in Japan. When disputes concerning legal rights and obligations or legal relationships arise, under modern legal systems, individuals are prohibited from resorting to force to recover a right that has been infringed upon by the other party (prohibition on self-help). Therefore, in order to resolve this sort of dispute in a peaceful manner, it is important that the government guarantees the right of access to the courts and establishes a litigation system.

Nevertheless, if the established litigation system is biased, said system would not properly guarantee citizens' rights, rendering it useless. Thus, in modern Constitutional States (or Rechtsstaat), it is necessary to establish a judiciary system in which the independence of the judiciary, that consists of the independence of judges and independence of the courts, is respected, and judicial decisions are made in public trials that any person is allowed to attend. This is why the Constitution of Japan vests all the judicial powers in the courts and prescribes that all judges shall be independent in the exercise of their duties (art. 76) and that trials shall be conducted and judgment declared publicly (art. 82).

Note, however, that in recent years, Alternative Dispute Resolution Procedures (ADR), which aim to resolve civil disputes by inserting a neutral third party in the dispute, have been established in order to help parties that wish to resolve civil disputes without resorting to litigation (established by the Act on Promotion of Use of Alternative Dispute Resolution of 2004). This is a system for resolving disputes without resorting to civil litigation when negotiations between the parties have failed. Conciliation and arbitration are some common ADR methods.

Development of related policies

When it comes to the Human Rights Infringement Cases (in a narrow sense, i.e. only those not related to constitutional matters) mentioned above, it is difficult to resort to the courts. Thus, in order to resolve them, one needs to resort to the operations of the Legal Affairs Bureau, Human Rights Commissioner, etc. The most common 'assistance' measures used by the Legal Affairs Bureau are the introduction of related administrative organs or related public/private entities, the introduction of legal aid services and the provision of legal advice and other measures found to be reasonable in order to help the victim, etc. (art. 13 of the above-mentioned Ordinance on Investigation and Resolution of Human Rights Infringement Cases). If we look at the 2015 statistics, we can see that among the total number of human rights infringement cases handled, these measures were taken in 92.1% of the cases. In contrast, more intrusive measures, as 'Explanation' (in which the person who has committed the infringement is persuaded to admit their mistake and take the necessary reparation measures), are rarely taken. Even if we add the measure called 'Demand', which demands that persons that are capable of taking effective measures to prevent or relieve the damage caused by the human rights infringement take the necessary measures, this accounts for only 3.5% of the total number of human rights infringement cases.

Regarding this point, we should pay attention to the policies related to human rights protection that the government has been implementing gradually in recent years. One is the Act on Promotion of Human Rights Policies (enacted in 1995 and consisting of a total of five articles). However, this act was a temporary law (or temporary-effect legislation) and lost its effect in March 2002. After that, it was replaced by the Act on Promotion of Human Rights Education and Enlightenment (shortened name). In fact, this only occurred after the 'United Nations Decade for Human Rights Education' was declared in a UN resolution, when the Japanese government decided to take domestic measures and formulated an action plan.

First, the government established the Human Rights Promotion Council and then, based on the achievements of the Council, established the Act on Promotion of Human Rights Education (2000; the official name is Act on the Promotion of Human Rights Education and Human Rights Enlightenment) in the form of a bill submitted by a Diet member. For this reason, a basic plan on human rights enlightenment was also approved by cabinet decision.

What this means precisely is that human rights is not specified in this human rights protection legislation. Nevertheless, the Basic Human Rights Education and Enlightenment Plan adopted by cabinet decision in 2002 defines human rights as 'inalienable rights that every person has based on human dignity, rights essential for guaranteeing the lives and freedoms of all members of society and essential for living a happy life in society'. As mentioned above, the public power is not specified as one of the parties in this definition, thus, human rights infringements may occur inside houses and private schools, and are extended to account for a much wider range. For example, the paparazzi phenomenon, when a large number of people gather around a perpetrator or a victim, is also regarded as a human rights matter.

Three human rights bills that contained these sorts of concrete measures for protecting human rights were submitted to the National Diet in the ten years between 2002 and 2012[7]. These bills were basically designed to give coercive powers to the Directive of the Ministry of Justice (Ordinance on Investigation and Resolution of Human Rights Infringement Cases), which as mentioned above does not have such powers. Nevertheless, all were discarded by the Diet. In other words, the main purpose of these bills was to establish a Human Rights Council inside the Ministry of Justice as an Independent Administrative Council and allow its members and employees to conduct compulsory investigations in the form of on-site investigations, inspections, enquiries, etc., that if not obeyed, would result in punishment.

Indeed, this tendency of the government may be explained by the above UN resolution concerning human rights education. However, what I want to point out here is that the 'human rights' protection that is being advocated in this context is different from the traditional constitutional understanding of fundamental rights or the violation or infringement of fundamental human rights. The same tendency may be seen in the educational environment. One example is the Study Group on Approaches to Teaching Human Rights Education established by the Ministry of Education in 2004 that prepared and published a total of three reports on Approaches to Teaching Human Rights Education between 2004 and 2008[8].

Either way, by looking at the contents of consecutive human rights protection bills, it is clear that there is a demand to engage in handling issues that, traditionally, have not been directly dealt with by constitutional law. Therefore, it is time for constitutional law to reconsider its understanding of human rights, or rethink the meaning of giving special treatment to the part of the law that regulates our relationship with the public power and is provided for by the Constitution.

Conclusion

Development of pluralism

As mentioned at the beginning of this chapter, ideals of or hope for a world without conflicts and disputes do not solve the problem. On the contrary, this form of idealism often leads to fundamentalism or the belief in the existence of an absolute truth. It is instead important to face the fact that we live in a world where there will always be conflicts and disputes, identify the causes, formulate efficient resolution systems and effectively implement said systems in order to solve them.

The model that seeks the peaceful coexistence of a variety of cultural groups while maintaining their cultural identities, i.e. the pluralist model, is one of the fundamental perspectives in terms of overcoming conflicts. When applied to the domestic level of politics, it leads to legal/political solutions such as the decentralization of power and the recognition of local autonomy or local self-government, etc. As mentioned above, these solutions have already proved effective in the cases of England and Spain, which have experienced terrorism over decades.

Yet, notwithstanding this fact, we should note that there is also significant criticism of the doctrine of cultural pluralism. Given the criticism regarding the need to avoid cultural assimilation and assimilation policies, we are forced to call this demand for unification into question and instead support multiculturalism, which claims that different cultural groups should be treated as equals.

Other important issues

Although I was unable to discuss it here, when we think about the management of a state or the UN, we should note that the financial basis to support these operations is an important issue.

For example, in Japan, the total amount of the general account for the fiscal year 2016 was 96.7218 trillion yen, the total amount of government bonds for debt redemption/interest payment was 23.6121 trillion yen (24.4%) and the remaining was invested in general policies such as social security, public works, education and science, etc. The expenditure related to social security accounted for 55.3% (31.9738 trillion yen) of the general annual expenditure, excluding tax allocations to local governments, etc. Also, the balance of accumulated public bonds at the end of the same fiscal year reached 838 trillion yen, indicating that the value of debt per citizen (including children) and per a family of four was around 6.64 million and 26.56 million yen respectively[9].

The UN is also in a difficult financial situation. The UN needs a large amount of funds to carry out its global programs promoting peace, human rights and humanitarian laws, the environment and democracy and fighting terrorism. Nevertheless, according to recent data, the regular budget of the UN over 2014–2015 (two fiscal years) was only $5.53035 billion (based on the resolution of the General Assembly of December 27, 2013). This amount is sufficient to cover the expenses of the 411,000 UN employees working at the UN Secretariats around the world (e.g. New York, Geneva, etc.) and the operating costs of all the secretariats. The countries that contribute the most to the UN are, currently, the US (22.0%), Japan (10.83%), Germany (7.14%), France (5.59%), England (5.17%), China (5.14%) and Italy (4.44%), but many member states are in arrears in the payment of their financial contributions and as of April 30, 2014, the total amount of unpaid contributions had reached $461 million[10]. Given this situation, there is an urgent need for reform within the UN.

10 Poverty and Educational Inequality as a Contemporary Global Issue

Hiroshi Sowaki

Introduction

Poverty is one of the global issues that Human Survivability Studies needs to address, given that it is closely connected to other global problems. For example, societies with chronic poverty are prone to the occurrence of violence. Poor countries in the midst of armed conflict are likely to become hotbeds for international terrorism. In turn, armed conflict results in human casualties, the collapse of food production/supply systems, the breakdown of healthcare and education services and economic failure, all of which create further poverty. In addition, natural disasters and abnormal weather can lead to poverty, and conversely, poverty magnifies the damage caused by disasters.

In this chapter, the reason poverty remains a global issue is first discussed, followed by ways we can examine poverty and the role of education based on the philosophy of Amartya Sen. Concerning measures to alleviate educational inequality as measures against poverty, educational programs effective in this regard will be examined using examples from the US in the 1960s and from the 1980s onwards.

Why is poverty an issue?

Why is poverty an issue? The 2005 Human Development Report by the United Nations Development Programme (UNDP) listed the following five points:

1. People's sense of social justice and morality do not allow extreme deprivation.

2. A policy that increases services or income for the poor and disadvantaged generates greater welfare than one that increases services or income by an equivalent amount for the rich and privileged.
3. Extreme inequality is bad for economic growth and the long-term efficiency of society as a whole.
4. Extreme inequality weakens political legitimacy and undermines the development of democracy.
5. Extreme disparity in income, health and education undermines public policy goals for improving welfare.

Economist Toshiaki Tachibanaki (2006) listed issues triggered by poverty, including declining economic efficiency and increasing unemployment, crime and the social burden brought on by aiding the poor, as well as ethical issues. He also discussed the risks posed by crime and disaster, and the emerging issue of health inequality in US society.

In regard to the decline in economic efficiency, a 2015 analysis by the OECD (2015b) revealed that it is not only the bottom 10%, but up to the bottom 40% whose low income reduces individual educational opportunities and prevents skills development, thereby dragging down economic growth.

To organize the above points, not only does poverty go against people's sense of fairness and justice as an intolerable inequality, it also leads to increased rates of crime, and disparity in health and education outcomes. These not only impair individual happiness, social security and safety, but also generate an increase in social costs and decline in economic efficiency. In addition, poverty has the potential to lead to societal segmentation and political instability and to weaken democracy.

The International Covenant on Economic, Social and Cultural Rights stipulates 'the fundamental right of everyone to be free from hunger' and 'the right of everyone to an adequate standard of living for himself and his family, including adequate food, clothing and housing, and to the continuous improvement of living conditions'. Poverty is a state in which such human rights are violated.

The fact that children also have such human rights has been confirmed by the Convention on the Rights of the Child. The cycle of poverty created from the poverty of children leads to the entrenchment of social class, and is a significant issue that will not only generate inequalities in results but also inequalities in opportunities.

Based on this understanding, countries have been implementing welfare (i.e., livelihood assistance and social security), labor (i.e., minimum wage system) and residential policies (i.e., public housing) as measures against poverty. Additionally, tax policies such as progressive taxation and public works projects are sometimes undertaken as measures against poverty. Further, most of the assistance from international organizations and developed countries' aid agencies given to developing countries are measures against poverty (refer to Chapter Fourteen of this book). Concerning child poverty, each country has child and maternal welfare policies. International initiatives include those taken by the United Nations Children's Fund (UNICEF). NGOs also play a significant role in these fields.

Several possible conditions that allow these policies to function can be considered. However, what is crucial here is that citizens are able to broadly participate in inclusive economic and political institutions, as illustrated by Daron Acemoğlu and James Robinson (2012) using the last 300 years of global history.

Methods to perceive poverty and its relation to education

One of the World Bank's goals is to end extreme poverty, and it has been collecting and analyzing data to achieve this aim. The poverty rate, defined by the World Bank, is the percentage of people who live on less than US$1.90 per day[1]. According to its data, the poverty rate has decreased from 34.82% of the world's population in 1990 to 10.67% in 2013[2]. However, there are still 766 million people worldwide who live under this international poverty line. The Sustainable Development Goals (SDGs) adopted in the September 2015 UN Assembly set its first target as 'ending poverty in all its forms everywhere'.

Although the poverty line of a few dollars per day is an understandable index, the specific details of adequate living standard and conditions, as stipulated by the International Covenant on Economic, Social and Cultural Rights, are not only at the bare minimum to be free from hunger, but determined by the correlation between economic, social and cultural factors. As such, there are absolute and relative aspects of poverty.

Poverty not only involves economic and material aspects measurable through income and consumption[3], it can also be examined by including low educational and healthcare standards. For this reason, the UNDP developed the Human Development Index (HDI), a comprehensive poverty index that includes factors other than income and integrates three fundamental aspects: (1) the healthcare index (life expectancy at birth), (2) the education index (years of schooling) and (3) the income index (gross national income per capita). The UNDP reports the HDI in its annual Human Development Report.

The philosophies of Amartya Sen, a Nobel-prize winning economist whose influence has reached the disciplines of philosophy and politics, served as the propelling force behind the publication of the Human Development Report. According to Sen (1992), poverty does not refer to the simple shortfall in income and utility, but is the failure of functioning capability consisting of 'beings' and 'doings', such as being 'well-nourished', 'in good health', 'able to read and write' and 'taking part in the life of the community'. This 'capability' approach not only identifies the people in poverty and the causes behind poverty, but also leads to policy recommendations that society should adopt.

In a lecture given in 1990, Sen (1991) discussed the poverty issue as follows: 'Persistent undernutrition is partly a matter of insufficient food intake, but the problem cannot be dissociated from that of deprivation of health care and basic education'. In 2002, Sen explained that to understand the demands of human security, the responsibility to provide basic education to help satisfy the right to security must be central; this social obligation must not only be fulfilled by the state, but also rests on all institutions and agencies. The poverty issue is not solely related to food and economic issues, but is also an issue of education. In particular, the role that basic education plays in terms of eradicating poverty is significant.

Goal 4 of the Sustainable Development Goals (SDGs) adopted in 2015, the draft of which was endorsed in advance by the Incheon Declaration for Education 2030, is as follows: to 'Ensure inclusive and equitable quality education and promote lifelong learning opportunities for all'. Before SDG 4, Goal 2 of the Millennium Development Goals (MDGs) drafted in 2000 had been targeting the aim of achieving universal primary education by 2015. Although 'education for all' along with MDG 2 had been promoted mainly by UNESCO, after the primary school enrolment ratio was increased from 84% in 1999 to 93% (estimated) in 2015, nearly fifty-eight million children were not attending school in 2012, and progress in reducing this number has stalled. In particular, there is a delay in sub-Saharan countries (UNESCO 2015).

The situation where primary education has not completely spread across developing countries is an issue related to disparity with developed countries, as well internal inequality within a country. Further, even if primary education becomes universal, issues pertaining to its quality and inequality in terms of opportunities to advance to secondary and tertiary education will not necessarily be resolved.

The 'War on Poverty' and education

Poverty can still exist in an affluent society. In the US, which experienced unparalleled prosperity following World War II, economist John Kenneth Galbraith indicated in his *The Affluent Society* (1958) that although society's concern toward the poor declines as their numbers decrease, poverty is an issue that cannot be eradicated by the overall increase in earnings. In 1962, social activist and author Michael Harrington reported in *The Other America* that the number of poor had risen to forty to fifty million, with an 'invisible country' existing within the US. Such proclamations and accusations motivated Presidents John F. Kennedy and Lyndon Johnson into action, with the US Federal Government moving to implement drastic policies to tackle poverty and inequality in education.

The 'War on Poverty'

After succeeding President Kennedy's administration following his assassination in November 1963, President Johnson discussed measures against poverty as one of the most important policies of the government, declaring 'War on Poverty' in the State of the Union address in January 1964. In addition to striving for the establishment of the Civil Rights Act, a pending issue since the Kennedy Era, the Johnson administration newly established the Economic Opportunity Act. It also achieved the Food Stamp Act in 1964, which made the provision of food to the poor permanent.

In 1965, Johnson, who was re-elected as president in the same year, established the Appalachian Regional Development Act, a regional measure against poverty. In terms of medical welfare services, revision was made to the Social Security Act to establish the Medicare and Medicaid systems. Regarding education, the Elementary and Secondary Education Act (ESEA), which stipulated federal assistance, and the Higher Education Act, which stipulated federal scholarship systems for university students, were established.

The central focus of this series of anti-poverty measures was the Economic Opportunity Act. In the US, self-help has been traditionally respected. As such, policies that guarantee such opportunities are the most accepted. Instead of giving grants directly to the poor, the Act provides education and occupational training for youth in poverty so that they can support themselves. To do so, the Act strove to improve opportunities in employment, public health, residence and educational facilities in poor regions.

An important initiative launched by the Economic Opportunity Act was Project Head Start. This project involves giving federal subsidies to regional organizations that provide childcare services to toddlers aged between three and four years from low-income groups. In addition to providing assistance to these toddlers, who cannot be expected to undertake self-help, the project looks at shaping them to become self-sufficient autonomous individuals by helping them to adapt to school and society. It is built on the principle of parental participation in regional activities and heightens the function of households to encourage autonomy. Given that this philosophy was compatible with US society, it was expanded even after the 1980s, with the Early Head Start program newly established for zero to two year-olds in 1994. Presently, this initiative has one of the largest budgets within the entire federal government.

Compensatory education

The initiative that played a particularly significant role within Johnson's presidency was the Elementary and Secondary Education Act (ESEA) of 1965, which focused on guaranteeing educational opportunities to children from poor households. Although in the US education is under state jurisdiction, and in most states, education lies in the jurisdiction of local school districts, its extremely decentralized structure meant that the state and local governments emphasized traditional and conservative values. As such, it prevented the implementation of redistribution policies aimed at education opportunity equality. The ESEA changed this structure significantly and gave the federal government the authority to grant subsidies to state governments to achieve equal opportunities.

In terms of the above-mentioned Project Head Start and the ESEA, education conducted with the objective of substantially guaranteeing education opportunity equality and thereby resolving the issue of poverty is referred to as 'compensatory education'. Its framework is informed by theories on cultural deprivation and human capital.

Cultural deprivation theory is also called the 'culture of poverty' and was disseminated through a series of best-selling books, including *Five Families: Mexican Case Studies in the Culture of Poverty* (1959), by cultural anthropologist Oscar Lewis. As characteristics of this 'culture of poverty', Lewis listed:

1. A lack of effective participation in the major institutions of the larger society.
2. Minimum organization beyond the nuclear and extended family.
3. Absence of childhood as a prolonged and protected stage in the life cycle.
4. Strong individual feelings of marginality, helplessness, dependence and inferiority.

His theory was later named 'cultural deprivation', explaining that the poor are unable to escape from a vicious cycle as the culture of poverty reproduces itself, and leads to the need for compensating this deprived culture using diligence and positive life attitudes through education.

Although the theory of cultural deprivation greatly impacted policymaking, it was criticized for forcing people in poverty to assimilate to the values of the white middle class, and claiming the cause of poverty was the responsibility of the poor. However, external assistance is indispensable to enable people to engage in self-help. In addition, requiring language ability or literacy for employment is common in all societies regardless of the values of social classes.

The human capital theory perceives abilities, such as knowledge, skills and expertise to engage in production activities, as human capital. As such, it is an economic theory that holds the view that educational investment can increase the economic value of a person, similar to investment in physical capital (Schultz 1961). This was sometimes perceived as focusing on highly talented human resources. However, the human capital theory explains an aspect of the cause of poverty by clarifying that the delay in the development of vocational ability from lack of educational training leads to low wages. As such, it demonstrates that human investment can serve as a correctional measure for poverty and inequality in education.

Equality of results

The Civil Rights Law (1964) required a survey to be conducted concerning the lack of availability of equal educational opportunities in US public schools along racial lines. James Coleman, a sociologist who supervised the survey, conceptualized five forms of educational opportunity equality when designing the survey (1968). Types 1 to 3 are approaches that have existed previously and involve equality in input, including school facilities and equipment, curricula and teachers. Coleman proposed two new approaches, Types 4 and 5, with Type 4 the equality of educational results between children of equal backgrounds (i.e., household socioeconomic status), and Type 5 equality in the educational results of children from unequal backgrounds (i.e., households that use languages other than English).

The equality of results has been criticized for taking away people's will and aspirations, mostly because it is misunderstood as uniformity in treatment. However, equality in educational results essentially refers to whether or not education has the capacity to produce similar results from children with similar backgrounds (i.e., household background). Further, this concept questions whether a similar result can be produced from children from less privileged environments by helping them overcome such environmental challenges.

The Equality of Educational Opportunity Study was conducted by extracting data from 645,000 schoolchildren from 3,100 schools across the US. Coleman intended to reveal the inequality between black schools and white schools within the public school system and to propose large-scale inequality correctional measures. However, the report released in 1966 showed that the gap between black and white schools concerning educational conditions, contrary to the prediction, was smaller when compared to regional differences. Instead, the report showed that educational outcome (student achievement) was most strongly affected by family background. In terms of school factors, the characteristics of peers, teachers and schools had an impact in this order. However, all these school factors were much smaller compared to family background.

Since the 1970s, the War on Poverty has been perceived as a failure or far below expectations, recently compounded by the financial crisis. As such, the public has become disenchanted with respect to improving society through social engineering methods. The Coleman study showed that the role schools play in closing the gap was small; this was subsequently used as a basis in arguments that attempted to reduce the government's educational investment.

However, as Thomas Piketty's *Capital in the Twenty-First Century* shows, the income gap in the US during the 1950s–70s was smaller than at any other time. Although social policies, including the 1960s education policies, may not have been able to eradicate poverty, they could be credited for having prevented the gap from widening.

Re-expanding educational inequality

In 1983, when deficits in American public finance and trade increased significantly, the Republican Reagan administration released the report called *A Nation at Risk*, prepared by the US Department of Education. In this report even elementary and secondary education were requested to seek not only equality, but also excellence, from the perspective of international competitiveness. To address the criticism that educational results do not correspond to federal and state governments' expenditure on public education, in the 1990s each state legislated stricter curricular requirements and held scholastic ability tests. State governments also emphasized schools' accountability by giving parents school choice and introducing private management to public schools.

The 1994 amendments to the Elementary and Secondary Education Act (ESEA) of 1965 in the Democratic Clinton administration made content and performance standards and student assessments mandatory for states. Further, the ESEA was revised during the Republican Bush administration in 2002, renamed the No Child Left Behind Act (NCLB). The official name of this law was 'An act to close the achievement gap with accountability, flexibility, and choice, so that no child is left behind'. As such, it clearly defines its objective of reducing the gap in scholastic ability. It also demanded that schools maintain strict accountability, including restructuring such as replacing school staff and closing schools.

The Act demanded that each state ensure each student group (i.e., economically disadvantaged students, students from major racial and ethnic groups, students with disabilities and students with limited English proficiency) meet the achievement goal by 2014. This was arguably an extremely ambitious goal that attempted to realize 'equality in results', as stipulated in Coleman's fifth type of equal opportunity. However, this goal was difficult to achieve in reality, with most states being granted waivers by utilizing alternative measures as a result. Although bipartisan, the law was criticized by conservatives who argued that the involvement of the federal government should be reduced, and even devolved to state governments. In contrast, liberals argued that it was excessively punitive and that the Act paradoxically led to widening the gap. In December 2015, the Every Student Succeeds Act (ESSA) was established by revising the NCLB Act. It provides states with more flexibility in establishing accountability measures, allowing schools to reduce reliance on testing while requiring them to give substantial weight to other academic indicators.

A comparison of the gap in scholastic abilities between the 1960s and the present shows that, despite the fact that the gap was reduced for a period, it has begun a trend of increasing since around 1990. The national student assessment test in the US aggregates according to race, showing that the difference in reading ability between black and white students aged thirteen decreased from thirty-nine points in 1971 to eighteen points in 1988. However, in 2012 it rebounded to twenty-three points. The score gap for mathematics also declined from forty-six points in 1973 to twenty-four points in 1986, but increased to twenty-eight points in 2012 (Snyder et al. 2016: 278, 291). Such an expansion of disparity in education has been generated in association with the expansion of the economic gap[4].

From the 1980s onward, reformation that further enforced the idea of self-responsibility was observed even in the area of welfare. These reforms were based on the approach of 'welfare to work' or 'workfare', and policies that promoted employment to enable welfare recipients to break free from the state of reliance on welfare. The most fundamental aspect of the welfare reforms in 1996 under the Clinton administration was the work requirements for individuals. For example, with temporary assistance for needy families with children (including pregnant mothers), people were required to participate in certain work activities no later than two years after receiving assistance.

As individual accountability was enforced and work was emphasized in welfare, the role played by education became more significant. This in turn required the enforcement of school accountability. Nonetheless, this approach has not necessarily led to the closing of the gap.

Educational programs as effective measures against poverty

Since the Coleman Report, it has been widely believed that schools have not invested much effect in correcting inequality. Rather, the reproduction theory that claims that schools functionally reproduce such inequality came to be advocated. The reproduction theory is a caveat against unconditionally relying on schools' capacity to educate. It is also significant in confirming the necessity of focusing on the entire social structure. However, this is unlikely to lead to gradual social reform.

Discussion on the effect of education has progressed, marked by the emergence of the view that sees the impact of schools as not necessarily insignificant. International comparative studies have showed that the effect of schools is larger in developing countries than in developed ones (Gamoran and Long 2007). In the former, which require the expansion of the quantity of education, the effect schools have is strong. In contrast, it has become harder to detect the effect of schools in developed countries' educational systems, which tend to have reached the stage of needing quality enhancement.

Further, research has been conducted on the effects of educational programs within the US, with results indicating that the following were effective measures against poverty: pre-school education, small classes, mentoring programs, curriculum reorganization, teacher training, a wage increase for teachers, scholarships for university students and intensive occupational training (Levine and Zimmerman (eds) 2010). In this section we discuss pre-school education and small classes, as well as some policies in developing countries.

Pre-school educational program

The Perry Preschool Project was the first research that validated the effect of high-quality pre-school education for the poor using RCT (see Chapter Five of this book). In this project, between 1962 and 1965, 123 high-risk, low-income black children aged three to five years at Perry

Elementary School in Ypsilanti, Michigan, were allocated randomly into either the group that received pre-school education or the group that did not. Follow-up studies were conducted continuously until today. The result when participants turned forty showed a significant difference in areas such as senior high school graduation rate, annual income and criminal record. Cost-benefit analysis was also conducted, reporting that for every $1 of expenditure spent on the program, a benefit of $12.90 accrued from tax revenue and a reduction in welfare costs, as well as savings in counter-crime measures and other areas (Schweinhart et al. 2005). As such, it is a desirable investment not only for those receiving pre-school education but also for the whole of society.

Similar results have also been seen in the RCT of the Carolina Abecedarian Project conducted from 1972, and the research on Chicago CPC (Child-Parent Centers) that commenced in 1985. With Heckman's statements that advocated for the effects of investment in pre-school education based on such factors as Perry's results (Heckman 2006) and OECD reports (2001, 2006, 2012, 2015a), such initiatives greatly influenced policies in the US and various other countries.

Nonetheless, according to the 2010 final report on the large-scale RCT that the Department of Health and Human Services launched in 1998, no long-term effects were detected from Project Head Start (US Department of Health and Human Services, Administration for Children and Families 2010). The 2013 report on the Tennessee Voluntary Pre-K Program (commenced in 2009) showed that a negative effect on cognitive capacity was detected at the time of completion of the first grade of elementary school (Lipsey et al. 2013). In other words, the results are mixed. Concerning this point, Perry stressed that it was effective for programs of a higher quality than Project Head Start (Schweinhart et al. 2005). Although the Tennessee Program revealed that the effects on cognitive ability disappeared two years after the program's completion, it did not negate Perry's results that revealed the effects of non-cognitive aspects. However, further validation using various perspectives and methods is necessary.

Class size reduction program for lower grades

During 1985–89, the State of Tennessee randomly allocated approximately 12,000 five-year-old pupils into either a small class of thirteen to seventeen students or a normal-sized class of twenty-two to twenty-five. This arrangement was sustained for four years, with a follow-up study conducted thereafter. As such a large-scale experiment requires funding decisions to be made at the state level, there have been no other examples. Therefore, such a study needs to be understood in conjunction with other smaller-scale experiments and statistical analyses. However, the study found that the class size reduction program was effective in lower grades, and in particular, more broadly effective amongst the poor segment of the population, and found that the effect was sustained in the long-term. Such outcomes had an impact on the class size reduction policies of each state and the federal government (United States Department of Education 1998). Angrist and Lavy (1999) showed that a similar level of effect existed in the above policies using regression discontinuity design (Angrist and Pischke 2009).

However, reducing class sizes requires increasing the number of teachers, resulting in increased financial burden and lower cost effectiveness. Thus, debate continues as to whether other policy measures are more effective.

Programs in developing countries

In developing countries, many RCTs have been conducted to prove the effectiveness of programs (Banerjee and Duflo 2011). Two of them are introduced in this subsection.

An RCT conducted in Kenya between 1998 and 2002 did not reveal any effect of the distribution of vermicides on school children's test score results. However, their health status improved significantly as did their attendance. Pratham, an NPO that provides quality education to poor children in India, is involved in sending young women as tutors for remedial classes. According to the program evaluation using an RCT conducted between 2001 and 2004, mathematics and language test scores increased, with the effect especially large for the former. These two examples are of programs that have managed to achieve great results at low cost.

In Mexico, a conditional cash transfer program for poor families, known as Progresa (later renamed as Oportunidades and now Prospera), was commenced in 1997. The conditions for households with children to receive cash include ensuring that their children attend school and receive a vaccination. An RCT was conducted upon implementation, showing that this program in particular increased the middle-school enrolment rates by 3.5% to 5.8% for males and 7.2% to 9.3% for females. The conditional cash transfer program is widely employed in Latin America and based on its outcomes, it is being attempted in Africa and Asia as well.

Conclusion

School education may not be able to provide a dramatic effect on the issue of poverty. However, by employing an effective program to correct educational inequality, it can contribute to resolving the poverty issue. Pre-school education is of particular importance.

Horace Mann, who is referred to as the father of the public school system in the US and the first Secretary of the Massachusetts State Board of Education, stated, 'Education then, beyond all other devices of human origin, is a great equalizer of the conditions of men – the balance wheel of the social machinery' (Mann 1848). In the US, where self-help is highly praised, public education is an important method for attempting to achieve equality and resolving the poverty issue. This was the case in the 1960s when social policies were expanded on a large scale, and the same is true in the current era of accountability.

In recent years, poverty and inequality, particularly the issue of child poverty, have become issues even in Japan. As early as 2006, the OECD first demonstrated in its *Economic Surveys: Japan 2006* that Japan had one of the highest relative poverty rates among OECD countries, with the child poverty rate rising to 14% in 2002, surpassing the OECD average. In addition, to ensure that poverty does not become entrenched, the OECD advocated the importance of ensuring that children in lower-income households have adequate access to high-quality education. In 2013, the Japanese Diet established the Law on Measures to Counter Child Poverty, stipulating assistance toward education, livelihood and the employment of guardians as necessary measures.

The 2011 OECD report (OECD 2011a) revealed that inequality has been on the rise in most OECD countries since the 1980s, and noted that facilitating and encouraging access to employment is a key policy challenge. As such, the report highlighted the importance of policies that invest in human capital for the workforce including education.

The role that education plays in the issue of poverty, a global issue, is significant. At the core of this approach is equal educational opportunity. To address poverty and educational inequality, one must carefully dismantle society's complex structures to explore the solution. To do so, comprehensive initiatives that encompass various fields are required.

11 The Threat of Infectious Disease to Humans

Masao Mitsuyama

Introduction

Human beings have confronted various threats with the potential to endanger their lives since their birth on Earth. Aside from the extinction of animal species caused by extreme changes in the Earth's environment, such as the glacial period, infectious diseases (contagious diseases) have been one of the most significant life-threatening factors affecting the human species. Cases can be verified through written records and oral histories.

Bacteria, which are classified into the prokaryote domain, appeared on Earth approximately three billion years ago, much earlier than the first eukaryotes (plants and animals) were generated. On the other hand, plants and animals are classified into the eukaryote domain. Originally, bacteria reproduced themselves by a simple mechanism of binary fission (dissociation into two daughter cells) in the natural environment, and gradually began to utilize plants and animals as seedbeds. They caused diversified and highly lethal diseases to living hosts. Viruses may have appeared before eukaryotes emerged. Viruses depend exclusively on host cells for their reproduction; they release their genome and literally hijack, or take over, the host cell's functions, such as replication mechanisms and protein synthesis. While the same infectious virions (virus particles) are reproduced, the host's cells are continuously destroyed.

Human beings have had no way of escaping from the attack (infection) of etiologic agents since their emergence. Records and historical evidence even reveal the possibility of human extinction. Viral diseases such as smallpox, influenza and measles, and also bacterial diseases such as the plague (black death) and cholera, have spread across the world as pandemics over the last 2,000 years. The outbreak of plague began in the Eastern Roman Empire in the sixth century AD and became a widespread contagion across Europe in the fourteenth century. The estimated number of victims was thirty million, roughly one-third of Europe's population at the time. A pandemic of recent times, influenza (called Spanish Flu), has been a problem since 1918; one-third of the population of the world was infected and approximately fifty million people were killed over the course of a few years. Another contagious disease, cholera, originally an endemic disease in the Bengal region, came to Japan. A cholera outbreak occurred in Calcutta (Kolkata) at the beginning of the nineteenth century, and the third wave, the third pandemic, arrived in Japan, claiming the lives of several hundred thousand people in Edo (Tokyo), one of the largest cities in the world. However, cholera is not of concern to most people these days, unless they visit a tropical or subtropical region where outbreaks continue to occur.

The terrors of the past must not be forgotten. As described above, pandemics caused by strong infectious agents were once unimaginably rampant in pre-modern times with poor medical facilities and treatment options. It is, however, important to remember that some dangerous infectious diseases continue to exist in some developing countries even today.

Robert Koch, the father of modern medicine, discovered the etiological agents for anthrax, cholera and tuberculosis, and the mechanism of infection was made clear in the latter part of the nineteenth century. It has only been about 130 years since the nature of bacteria and viruses as pathogenic microbes was clarified. However, the rapid progress of medical microbiology made people believe in the early 1970s that most infectious diseases were no longer a threat, at least in developed countries. Even though they were not completely eradicated, people were under the impression that infectious diseases were perfectly under control thanks to the development of vaccines and antibiotics. In fact, that perception that held sway in the 1970s is a misunderstanding. Emerging infectious diseases have broken out since the 1970s, and re-emerging diseases are also becoming a serious problem. Classical infections, regarded as diseases of the past, have reappeared (see Table 11.1). Why has this occurred? New microbes have been discovered, although it was believed that all had already been identified. Emerging and re-emerging diseases have occurred one after another over the past forty years, since the time when people believed that vaccines and antibiotics could control infectious diseases.

Table 11.1: Emerging infectious diseases

1973	*Rotavirus*	Virus	Infant diarrhea
1976	*Cryptosporidium parvum*	Parasite	Severe diarrhea
1977	*Ebolavirus*	Virus	Ebola hemorrhagic fever
1977	*Legionella pneumophila*	Bacterium	Legionnaires' disease
1977	Hantavirus	Virus	Hemorrhagic fever with renal syndrome
1977	*Campylobacter jejuni*	Bacterium	Diarrhea
1980	HTLV-1	Virus	Adult T cell leukaemia
1981	*Staphylococcus aureus* (TSST+)	Bacterium	Toxic shock syndrome
1982	*Escherichia coli* O157:H7	Bacterium	Hemorrhagic colitis
1982	*Borrelia burgdorferi*	Bacterium	Lyme disease
1983	HIV	Virus	Acquired immunodeficiency syndrome
1983	*Helicobacter pylori*	Bacterium	Gastric ulcer
1985	*Enterocytozoon bieneusi*	Parasite	Persisitent diarrhea
1986	*Cyclospora cayetanensis*	Parasite	Persisitent diarrhea
1988	HHV6	Virus	Exanthema subtum
1988	Hepatitis E virus	Virus	Hepatitis E
1989	*Ehrlichia chafeensis*	Bacterium	Ehrlichiosis
1989	Hepatitis C virus	Virus	Hepatitis C
1991	*Guanarito virus*	Virus	Venezuela hemorrhagic fever
1991	*Encephalitozoon hellem*	Parasite	Conjunctivitis
1992	*Vibrio cholerae* O139	Bacterium	New type Cholera
1992	*Bartonella henselae*	Bacterium	Cat scratch disease
1994	BSE prion	Prion protein	New variant CJD
1994	*Sabia virus*	Virus	Brazilian hemorrhagic fever
1995	Hepatitis G virus	Virus	Hepatitis G
1997	Avian Influenza virus (H5N1)	Virus	New type Influenza
1999	*Nipah virus*	Virus	Encephalitis
2003	SARS CoV	Virus	Severe acute respiratory syndrome
2009	Swine Influenza virus (H1N1)	Virus	New type Influenza

This chapter presents, drawing on historical material, a general view of pandemics that circumvent quarantines and spread across borders, and examines the need for Human Survivability Studies (HSS) to deal with this issue.

Lessons from the history and eradication of smallpox

Smallpox (variola) must be identified as the most disastrous threat among infectious diseases in historical accounts. This disease, with very high transmission and fatality rates, is caused by poxvirus infection, also known as variola virus. Smallpox symptoms include a high fever and maculopapular rash all over the body. No credible description is available regarding the area of emergence of variola, but a recent investigation revealed that the mummy of Ramses V, the Egyptian king around 1100 BC, showed a systemic scar caused by the characteristic maculopapular rash. Smallpox did not exist in the Americas prior to Columbus' exploration in the late fifteenth century. The virus was carried there by people and domestic animals such as cows, horses and honeybees, and with food such as wheat, and rapidly spread among Native American populations. Historical records from Japan tell that many members of the Fujiwara family died due to smallpox in the eighth century. Even Emperor Go-Daigo was diagnosed with the disease in the tenth century. Three widespread outbreaks occurred in the Meiji era; the number of deaths in the first outbreak was 32,000, with 24,000 in the second and 16,000 in the third wave.

People experienced and realized, even back then, that those who were infected with smallpox and fortunately survived did not become re-infected during subsequent outbreaks. This is 'immunity'. In the ancient history of Turkey, empirical variolation (inoculation of a specimen from human smallpox) was performed for prevention. Regarding prevention methods, a Japanese physician, Ogata Shunsaku, performed intranasal variolation during an outbreak in the Chikuzen Province in Kyushu in 1790. This was a pioneering trial in Japan, but it involved the risk of death of the vaccinee. In 1796, Edward Jenner developed a safe vaccination using cowpox that caused rather mild local lesions at the site of inoculation. The procedure was conveyed to Japan via the Dutch East India Company during the period of national seclusion, and Koan Ogata, one of the pioneer physicians, established Joto-kan (Smallpox Vaccination House) in Osaka, Japan in 1849. This safe vaccination was gradually accepted and rapidly spread all over the world (the cowpox vaccination was called *hakushin*, meaning white god during the Edo era in Japan). As increasing numbers of people were vaccinated, those with immunity increased in number and the victims of smallpox gradually decreased. However, smallpox itself could not be eradicated. Progress was made through continuous efforts. Attenuated vaccinia viruses replaced cowpox viruses. The effectiveness of the vaccine was certified and the World Health Organization (WHO) passed a resolution calling for the global eradication of smallpox in 1958 by worldwide extensive vaccinations. The prevalence and infection of smallpox were not recognized at any place in the world in 1977, and the complete eradication of the disease was officially declared by the WHO in 1980. Finally, humans are now free from the menace of smallpox.

This was a victory for modern medicine. However, excluding smallpox, little progress has since been made toward achieving complete eradication of infectious diseases. The reasons smallpox was able to be eradicated are as follows.

1. Variola has no animal hosts aside from humans; variola is a human-prone infectious disease because of the presence of strict receptors for this virus.
2. Transmission of the virus causes overt clinical symptoms without exception, and there are no latent or chronic infections.
3. It is easy to identify infected people because they show the typical symptoms of a high fever and maculopapular rash all over their body.
4. Antigenic mutations are not seen in the variola virus. Thus, vaccine effects are stable.
5. The effect of vaccination is nearly 100%.

Thanks to these factors, variola (small pox) has been eradicated. Conversely, it is unfortunate that other infectious viruses are capable of infecting various animal hosts, including bacteria, in the environment and in nature. Furthermore, antigenicity tends to change in the case of viruses due to spontaneous mutation and selection. Thus, it is extremely difficult to create a powerful vaccine like the cowpox vaccine. The global polio eradication project has proceeded and its success is expected soon. It is, however, theoretically impossible to eradicate infectious diseases other than variola, in spite of the fact that some vaccines are able to prevent them to some extent.

Influenza and the problem of virus mutation

Influenza outbreaks are a problem every year, and avian influenza is especially concerning. The horrible memory of the Spanish Flu, which humans confronted approximately 100 years ago, still remains.

In Italy in the Medieval Age, astrologists hypothesized that influenza occurred when the stars in the sky showed extraordinary movement. Later, the disease was named after the 'influence of celestial bodies' in England. Influenza is a systemic disease caused by viral infection, completely different from the common cold accompanied by nasal and upper respiratory inflammation. In Japan, the disease was called *in-furyu-enza* during the end of the Edo era.

Shiwabukiyami, meaning 'disease with cough' in Japanese, appeared in written records in Japan around the tenth century, at exactly the time of influenza in the modern understanding. Influenza prevailed repeatedly after the middle of the Edo era and the outbreaks were named *Tani-kaze* in 1784, *O-shichi-kaze* in 1802, *Tsugaru-kaze* in 1828 and *O-some-kaze* in 1891. The Spanish Flu pandemic swept across the world in the most hazardous proliferation of influenza from 1918 to 1919, before the influenza virus was isolated and identified. It has been claimed that the earliest known cases struck at a military post in Kansas, US, in March of 1918, and the flu swiftly spread across that country. In June, patients were seen in Western Europe, and 30% of the population of Spain was infected. The first wave of the 1918 pandemic swiftly spread from the spring to the summer. The second wave, which erupted in the autumn, was much deadlier with a higher virulence. Infections were seen among the young, and the mortality rate of this wave increased to ten times as high as that of the first wave. In the winter, Spanish Flu expanded into a third wave. The deadly Spanish Flu virus, according to some descriptions, attacked as many as 500 million people worldwide, and this record-breaking pandemic killed forty to fifty million people. There were twenty-three million infected people in Japan, and approximately 400,000 died.

In the case of smallpox, people gained lifetime immunity as they recovered from the variola virus, while in the case of influenza, people were repeatedly infected. How is that possible? The RNA genome in the influenza A virus is more likely to mutate. In addition, the genome of the influenza A virus is comprised of eight segments, which is exceptional for a virus genome. Two essential molecules, hemagglutinin (HA) and neuraminidase (NA) are coded by the segmented genome to infect a host, stick to and invade respiratory cells, release genomes and replicate and proliferate. Then, proliferated virions leave the host cells and expand into neighboring cells. It is highly probable that in the host cells, different types of HA(H) and NA(N) genes are likely to be incorporated into virions during assembly. As a result, even if antibodies specific for a particular type of H antigen or N antigen are generated, the immune defense of the antibody has no effect at all against viruses of different H/N serotypes. In fact, the virus responsible for the Spanish Flu was the H1N1 serotype. As for subsequent influenza pandemics, Asian Flu (1957–1958) was caused by a strain of H2N2 (an estimated one million people were killed worldwide) and Hong Kong Flu (believed to have started in 1968) was caused by a strain of H3N2 (an estimated 500,000 people died worldwide). Russian Flu was a regional 'benign' pandemic in 1977. This form of influenza was caused by a strain of H1N1, the same serotype as Spanish Flu.

When a pandemic occurs, many affected people gain immunity to that serotype and infection and transmission are less likely to occur, even if the virus exists in the environment. Another pandemic would occur if people who acquired immunity were attenuated due to the lapse of time or if a new serotype of virus were encountered. The population will be infected without resistance against a new strain, and will have no way to avoid it. A new antigen with discontinuous variations could cause another pandemic. As far as influenza A virus is concerned, the existence of H1–H16 and N1–N9 is known. Further, there are an extremely large number of potential combinations. Originally, viruses infected specific host animal cells in nature. The host specificity would determine the host range. In other words, humans are not infected by the viruses that are specific to a certain animal host.

In the past it was thought that species barriers existed. However, presently, new and serious patterns of pandemics have the potential to occur. Viruses now mutate, although, for example, some viruses used to select only pigs and birds (poultry) as specific hosts in the past. Thus, it is possible for non-human-specific influenza viruses to infect humans. Actually, the H5N1 type avian influenza virus was detected as infectious to humans in Hong Kong in 1997. The mortality rate was extraordinarily high, and five of eighteen patients were killed, but the scale of prevalence was not large. People were shocked at the first experience of a new pattern of the infectious disease. The outbreak of swine flu in Mexico in 2009 is still fresh in our memory. The disease started to prevail in Mexico, with 120,000 people infected and 10,000 deceased.

What all this suggests is that the key is the ability to mutate. The vigorous mutation of influenza viruses enables repetition of the pandemic. Numerous animals on Earth have specific virus forms. Aquatic migratory birds move in a borderless world and their travels contribute to the spread and expansion of new types of viruses. Moreover, raising pigs is common in both developed and developing countries, and pig farms can be located close to poultry farms. Pigs are regarded as a 'mixing vessel' for the development of new mutant viruses with mixed genomes. As described above, deeply analyzing the background factors to the spread of viruses, we can state the following.

1. Mutation and recombination of viral genes are viral factors.
2. Animals enable genome mixing and mutation.
3. Migratory birds are transmission devices.
4. Declining herd immunity occurs due to time lapse.

It must be recognized that in terms of the spread of viruses, extremely complicated and essential factors exist in a complex context.

Cholera pandemic: History and factors for expansion

Cholera was originally endemic in the Indian subcontinent. However, the disease, initially in a limited location, expanded worldwide at the beginning of the 1800s as the first pandemic began. Following the first pandemic, six pandemics occurred successively. The seventh pandemic is ongoing. Cholera, which was endemic in the Bengal bay region, with causative bacterium *Vibrio cholerae* inhabiting the sea, became a worldwide pandemic in modern times. One of the mechanisms behind the spread of cholera is sea-routes. Ship transportation has drastically developed since the medieval ages, especially since the modern ages, as increasing numbers of people came to travel by sea. Once *Vibrio cholerae* has invaded and infected someone, it is accompanied by symptoms including watery diarrhea, triggering tremendous amounts of contamination. In cases of severe symptoms, more than ten liters of diarrhea can be excreted. More than 100 million cells of *Vibrio cholerae* exist in the stool of affected patients, in the condition of axenic cultures per one milliliter of watery diarrhea. In this manner, *Vibrio cholerae* contaminate the water and surroundings. Before the causative agent was isolated and identified, John Snow, a London-based physician, found a link between cholera and contaminated drinking water. When the cholera pandemic occurred in England in 1854, Snow found that city wells in the area of Broad Street were contaminated, after tracking down the places where patients emerged and places where drinking water was provided via wells and plotting these locations on a map. Then, Snow had some of the wells blocked off and eventually succeeded in halting further rampancy of cholera. Snow's study became well known as a significant model for leading advances in the field of public health and epidemiology.

Japan suffered a cholera outbreak around 1822, when the first pandemic was carried into the Kyushu district, probably from the Korean peninsula, and was dispersed to Western Japan. Following that, in 1858, another pandemic erupted and cholera claimed hundreds of thousands of lives in Edo. Historical records even reveal there was a shortage of coffins. Cholera was called *korouri* in association with the pronunciation of *korori*, an adverb meaning 'easily and suddenly.' As this word indicates, after infection people were killed instantly by dehydration after severe diarrhea caused them to lose massive amounts of body fluid. Nevertheless, the disease started to degrade at the beginning of the Showa era. Except for the temporary prevalence of cholera among repatriates at the end of World War II, cholera has not spread within Japan since 1970, owing to changes in pathogenic bacterium. Classical cholera with very high pathogenicity was replaced around 1960 by El Tor cholera with milder virulence.

The pathogenicity of cholera has been microbiologically analyzed, and the mechanism of colonization and the structure of the toxin protein have been understood to a great extent. Methods of prevention have been developed, such as vaccines for travelers in endemic areas (putting effectiveness aside) and antimicrobial agents. In spite of these efforts, cholera has not been eradicated because of the following reasons.

1. The sea is the original habitat of *Vibrio cholerae*.
2. Many *Vibrio cholerae* are created in the condition of a pure culture. Water-like diarrhea excreted by patients contains so many living bacteria that environmental contamination cannot be avoided.
3. *Vibrio cholerae* can survive even at very cold temperatures, and frozen seafood made in Southeast Asia is exported all over the world.
4. There used to be an inaccurate idea regarding the sustainability of bacterial life. It has been elucidated that *Vibrio cholerae* in cold water exists in the condition of VBNC (viable but non-

culturable). This means that even if a colony is not detected by culturing the sample, bacteria can survive there and then start to proliferate after the environmental conditions become appropriate.

Considering these factors, prevention methods are as follows.

1. Heating food and water for drinking.
2. Segregating infected patients in the early stages.
3. Detecting sources of contamination as a matter of urgency.

The above principles are observed thoroughly during outbreaks today.

West Nile virus: A new virus strain invades the US

West Nile virus is of the family Flaviviridae, including the Japanese encephalitis virus and over seventy other closely related viruses. West Nile virus was first isolated in the West Nile district of Uganda in 1937. After the identification of the new virus, infected patients were identified in Israel, France and Eastern Europe. No infections were located in North America until July 1999, when an abnormal phenomenon was seen around the Bronx Zoo – numerous crows and other birds lost their sense of direction, fell down and died. In August, two humans were infected with encephalitis. The New York City Health Department investigated and reported six more cases of people with signs of encephalitis. Blood specimens (blood and bone marrow) of these eight people were analyzed by the Center for Disease Control and Prevention (CDC) in Atlanta. The CDC initially suggested a diagnosis of St. Louis encephalitis (an acute febrile infectious disease caused by flavivirus). This kind of encephalitis had been seen commonly in the US at that time. By the end of the summer, the symptoms had appeared in fifty-five people and six died. However, St. Louis encephalitis had never been recognized in New York prior to this incident. In addition, birds with St. Louis encephalitis had never died before.

Tracy McNamara, the head of the department of pathology at the Bronx Zoo, became concerned that a large number of crows, bald eagles and flamingos, which had been at the zoo, had died. McNamara also felt skeptical about the cause of the deaths and believed that the birds were not killed by St. Louis encephalitis. She therefore requested further investigation. Finally, the virus was proved not to be St. Louis encephalitis, but in fact West Nile virus.

West Nile virus, upon entering and being detected in New York, continued to spread. The next summer, in 2000, West Nile virus spread along the east coast and invaded eleven states, including Washington D.C.. It was proved that more than sixty species of birds were infected in that state. In the summer of 2001, West Nile virus was detected in wild birds in Ontario, Canada. In 2002, more than 1,400 people became ill after being infected, and 284 people died from the disease. Moreover, in 2003, West Nile virus was confirmed on the west coast of Canada, resulting in more than 9,000 sufferers and 231 deaths.

The amplifier hosts were birds and the vectors were house mosquitoes, *Culex* and *Aedes*. The incidence of West Nile fever is considered to appear in approximately 20% of those infected. According to statistics by the CDC, between the summer of 1999 and 2013, 39,557 cases of infection (17,463 encephalitis cases) and 1,668 (4%) cases of death were confirmed. The routes of invasion have not been exactly determined even today. It is highly probable that West Nile fever was carried to the US by wild birds. The disease could even have come from a mosquito carried on a plane. As for the domestic dissemination from the east to the west coast, it is most probable that infected mosquitoes were carried on domestic airplanes.

Learning from experience and retaining knowledge of sudden invasions and settlement in America, the lessons to be learned are as follows.

1. An endemic is not necessarily confined to a specific region and can easily spread to, and settle in other countries.
2. Even when the disease is not directly contracted from person to person, through vector mosquitoes (*Culex* and *Aedes* for West Nile fever), viruses can spread from a patient to a healthy person.
3. As long as vectors exist, diseases are inescapable and likely settle in a region (country) once viruses invade.

The above lessons are a reminder of dengue virus, which suddenly became a significant problem in Japan in the summer of 2014. Dengue fever erupted for the first time in seventy years in Japan, and the virus is one of the Flaviviridae family with the mosquito vector of *Aedes albopictus*. West Nile fever might indicate the future of dengue in Japan.

Emerging infections and drug-resistant bacteria

The term 'hospital-acquired infection' has become familiar via the mass media. Since hospitals are places to cure diseases, patients should not acquire another disease while staying at them. Unfortunately, recently, it increasingly happens that inpatients acquire fatal diseases such as fatal septicemia at university hospitals and medical institutions where highly advanced medicine is practiced, rather than at small-scale community hospitals. Most of the diseases are caused by methichillin-resistant *Staphylococcus aureus* (MRSA), multiple-drug-resistant *Pseudomonas aeruginosa* (MDRP), vancomycin-resistant *Entercocci* (VRE) and others, which are normal inhabitant bacteria that do not cause serious infection in normal cases. These kinds of bacteria take hold in hospital environments, such as in wards and on linen and medical staff, and they do not infect healthy medical staff. Those who are infected are compromised hosts, so to speak, whose original biological defense capability is lowered, such as inpatients with chronic diseases including diabetes, hepatic cirrhosis and others, those with a serious cancer, those who are administered immunosuppressive drugs and very elderly people whose host defense mechanism is impaired. In addition, the causative bacteria of hospital-acquired infections have genetically acquired a resistance against a variety of antibacterial drugs that would normally be effective against such bacteria. In other words, if there is no drug resistance, any infectious disease can be treated as far as we have effective antibacterial drugs. However, the acquisition of drug resistance has enabled commensal or weak bacteria to infect immunocompromised hosts and can result in untreatable diseases.

What should be recognized is that:

1. No matter how far medicine advances, it is impossible to completely remove the bacteria inhabiting the environment and body (on skin and in intestines).
2. Owing to the advancement of medicine, even hosts whose biological defense capability is extremely low can survive.
3. Many kinds of bacteria have special gene transmission mechanisms and also have the ability to introduce and acquire drug resistance genes from outside.
4. Even if an antibacterial drug effective to drug-resistant strains of bacteria is developed, bacteria with new resistance genes inevitably emerge.

If the conditions are right, bacteria divide and multiply within thirty minutes. If one out of 100,000 bacteria acquire resistance, even if 99,999 bacteria are eliminated with drug treatment, the drug-resistant bacterium that survived will multiply to a group of about 10,000 bacteria after eight hours. At present, when the development of an antibacterial drug is said to cost billions of dollars, in a modern society where there are too few pharmaceutical companies working toward the development of new drugs against infection, and in view of the present situation where advanced medical practice aimed at saving lives inevitably lowers humans' ability to protect against infection, it will be difficult to respond to the problem of infection in the near future. These kinds of infections are different from those examples mentioned above. However, the medicine of developing countries is coming close to that of developed countries. If the medicine of the level of developed countries, including mass administration of antimicrobial drugs, is implemented throughout the world, it is highly likely that opportunistic and hospital-acquired infection will occur everywhere.

Challenge of global infections and HSS

Because of space limitations, the following are brief examples to facilitate understanding of the background factors of infections. It has often been understood that, generally speaking, infections and communicable diseases are tasks that only medical staff, medical researchers and public health experts can deal with. First of all, infections and communicable diseases are deeply involved in very complicated and in common social backgrounds. In ancient times infections and communicable diseases were limited to special areas. However, they are rapidly beginning to show global trends. What should be emphasized, in particular, are the global movements of food, food products and humans and their speed in contemporary society. The Age of Exploration in the Middle Ages made trade across wide areas possible and spread technologies, and at the same time, it bore the fate of propagating infections. However, when their speed was slow, it was possible to anchor a ship offshore for a certain period and quarantine it. The original meaning of the term 'quarantine' is forty days, and it derives from the crews of ships who were allowed to land after it was confirmed that they were not infected by plague or other diseases after a period of forty days. According to the statistics of the Ministry of Land, Infrastructure, Transport and Tourism of Japan (MLIT), the number of passengers

Table 11.2: *Top 10 causes of death*

	Diseases	No. of deaths (x 1000)	% of total deaths
1.	Ischemic heart disease	7250	12.8
2.	Cerebrovascular disease	6150	10.8
3.	Lower respiratory tract infection	3460	6.1
4.	Chronic obstructive lung disease	3280	5.8
5.	Diarrheal disease	2460	4.3
6.	HIV/AIDS	1780	3.1
7.	Respiratory cancer	1390	2.4
8.	Tuberculosis	1340	2.4
9.	Diabetes mellitus	1260	2.2
10.	Traffic accident	1210	2.1

on international routes rose from eight million in 1975 to more than 60.3 million in 2012. In the modern age of large-scale production and consumption, tremendously large quantities of food and food products are exported all over the world by air. In addition, migratory birds that cannot be controlled continue to freely cross borders and carry pathogens.

Given global infections, and infections potentially worsened by the development of medicine in the future, the approaches required from the perspective of human survivability are the real-time collection of information, prompt analysis and prediction and the review of the isolation and containment of pathogens and the expanded artificial movement of infections by airplane.

Finally, let's take a look at the ten leading causes of death worldwide. Table 11.2 shows the ten leading causes and numbers of deaths prepared from 2011 WHO statistics. It is understandable that ischemic heart disease and stroke rank first and second. Out of the other diseases, lower respiratory infection, diarrheal infection, AIDS and tuberculosis cover almost all infections. The total number of deaths caused by these diseases, ranking third, fifth, sixth and eighth respectively, is 9.04 million and emerges as the top causes of death, overtaking ischemic heart disease which ranks first. Please take this as confirmation that the infectious diseases that are widely thought to be obsolete are the most serious diseases in the world.

12 The Population Problem and the Food Problem: Beyond the Limits of Individual Sciences

Koichiro Oshima

One of the challenges we face in the twenty-first century is the population problem. In October 2011, the world's population reached seven billion. According to UN estimates, the human population of the world is expected to continue growing to reach ten billion by around 2083. Humankind has increased its population by improving food, clothing and shelter through technological innovation. Can everyone live in affluent circumstances in the future? Can we produce enough food to nourish the whole population? The population problem is simultaneously a food problem.

Population growth and distribution

Population growth and the lengthening of the average lifespan are dependent on stable food production. It is estimated that 10,000 years ago the world's population was a few million. Agriculture emerged around that time, resulting in population growth acceleration. By 1 AD the human population reached 100–300 million, increasing to 200–400 million by 1000 AD. The rate of increase escalated further past 1700 AD, and by 1800 there were about one billion people. The global population reached two billion in 1930, three billion in 1960, four billion in 1974, five billion in 1987 and six billion in 1999. In this way, a rapid population increase, or 'explosion', has occurred in the space of a few hundred years.

The Japanese population increased significantly until the first half of the Edo period (1603–1868). In around 1500 AD, Japan had about twelve million people, rising to thirty-one million around 1720. The population growth leveled off in the second half of Edo for 150 years until the beginning of the Meiji period (1868–1912) when it resumed expanding. The population at the end of Meiji was four times as large as that of the first year of that period; it reached fifty million around 1910, and went beyond 100 million in 1967. At the same time, the average lifespan of Japanese people has also lengthened. Life expectancy, only thirty years in the seventeenth century, increased to fifty around 1940, seventy in 1964 and eighty-three (male 80.21, female 86.61) in 2014.

Where do the seven billion people live on this Earth? Four-point-two billion people, or about 60% of the world's population, live in the Asian region. Six countries amongst the ten most populated countries are in Asia: China is first (1.35 billion), India second (1.24 billion), Indonesia fourth (240 million), Pakistan sixth (180 million), Bangladesh eighth (150 million) and Japan tenth (130 million). Although the population in the Middle East and Africa is currently not that large, it is expected to account for a large proportion in the future, based on the high rate of increase.

Prediction of massive famine

In the late eighteenth century, 1798 to be precise, Thomas Robert Malthus published *An Essay on the Principle of Population*, in which he predicted that '(t)he power of population is so superior to the power of the earth to produce subsistence for man… gigantic inevitable famine stalks in the rear'. Fortunately, this prediction did not come true, because humankind was able to secure vast tracts of farmland through the development of machinery in the industrial revolution and the American expansion towards the western frontier, supported by such development. Also, at the end of the nineteenth century (1898), Sir William Crookes warned in his inaugural address as President of the British Association for the Advancement of Science, that all civilized nations stand in deadly peril of facing famine in thirty years. Crookes based his forecast on the sense of crisis that the world population was growing, while food production per head had begun to decrease. He also argued that if Chile saltpeter (potassium nitrate) – the basis of nitrogen fertilizer, indispensable for crop production – were to run out, it would become impossible to produce enough food to nourish all of humankind. He added that, in order to prevent this from happening, the development of a scientific method for transforming atmospheric nitrogen into fixed nitrogen was necessary. Two scientists, Fritz Haber and Carl Bosch, found an answer to this conundrum. They established the Haber-Bosch process ($N_2 + 3 H_2 \rightarrow 2 NH_3$), a method of converting atmospheric nitrogen into ammonia by a reaction with hydrogen. Sir William's prediction did not come true, thanks to this great invention. The Haber-Bosch process is still used around the world, pulling unlimited amounts of nitrogen from the air. It is this great invention that saves seven billion people from starvation.

Agrochemicals and environmental problems

Agricultural chemicals are used to eliminate plants or animals injurious to crops, such as pathogenic bacteria, insects, mites, rats and weeds. Although agrochemical-free cultivation or organic farming is praised, providing farm products without using agrochemicals requires a great amount of time and labor. In fact, it is said that food production would decrease by 30% if agrochemicals were not used. Conversely, environmental pollution caused by agrochemicals and health hazards resulting from residual pesticides are becoming significant problems. Insecticide, germicide and herbicide are the three main agrochemicals, but there are also others such as rodenticide and attractant. If we focus on agrochemicals' chemical structure, they can be categorized into several types, such as organophosphorus compounds, chlorinated organic compounds, carbamate and pyrethroid. Although organophosphorus agrochemicals were claimed to be of low toxicity and do little harm to human bodies, multiple cases of damage to human bodies occurred once such chemicals started to be used in 1955; more than 5,000 people lost their lives every year due to organophosphorus agrochemicals between 1955 and 1971. Today, the death rate from such chemicals is very low, thanks to people's increased understanding of and heightened interest in agrochemicals. Acute poisoning by organophosphorus agrochemicals occurs as they combine with an enzyme in the body, namely acetylcholine esterase, and inhibit neural transmission. Because of this, if muscles are stimulated and contract, they remain excited and cannot relax, losing their normal functions. This results in suffocation and paralysis of the limbs. Parathion, sumithion and malathion are typical organophosphorus agrochemicals.

Rachel Carson discussed several chlorinated organic agrochemicals, such as DDT, BHC and 2,4,5-T, in her book *Silent Spring* (1962), where she documented a composite of examples drawn from many communities where the use of DDT had caused damage to wildlife, birds, bees, livestock, domestic pets and even humans. Paul Muller

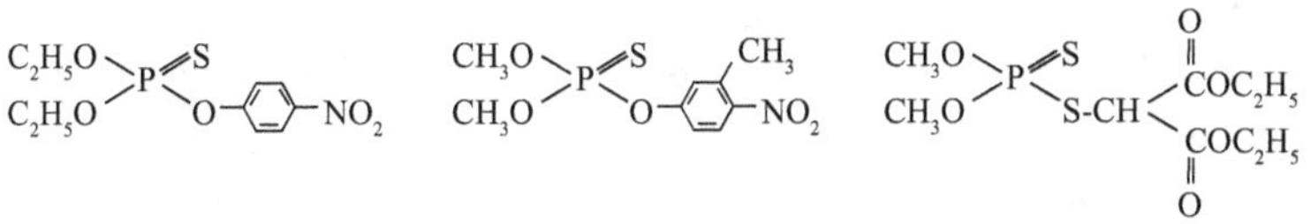

Parathion Sumithion Malathion

Figure 12.1: Structural formulae of parathion, sumithion and malathion

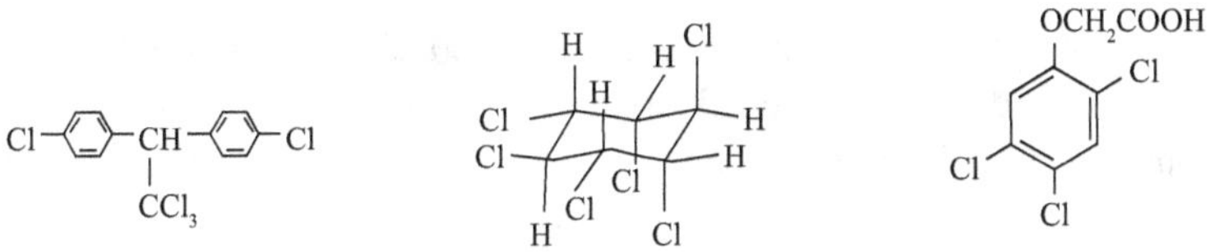

DDT BHC 2,4,5-T

Figure 12.2: Structural formulae of DDT, BHC and 2,4,5-T

received the Nobel Prize in Physiology or Medicine in 1948 for his discovery of DDT. It is effective in the control of petechial typhus and malaria, but its usage is restricted because of its function as a hormone-disrupting chemical. BHC production is prohibited worldwide today.

Origins of chlorinated organic agrochemicals

Compared with organophosphorus agrochemicals, chlorinated organic agrochemicals have less acute toxicity, but they are stronger insecticides and have longer residual effects. When chlorinated organic agrochemicals are sprayed, they spread to crops, soil, water and air, and concentrate in living organism throughout the food chain. Agrochemicals, released into rivers and the sea, eventually accumulate in human bodies through this biological concentration, starting from seafood. Chlorinated organic agrochemicals have similar insecticidal effects to organophosphorus agrochemicals, as they block normal neural transmission. The process in which such poisonous effects are generated, however, is different from organophosphorus agrochemicals. Chlorinated organic agrochemicals hinder neural transmission by impeding control of the concentration of sodium potassium ions inside and outside of neurons; in other words, they delay the closure of the sodium gate and thus keep neurons stimulated. This disrupts normal neural transmission, and results in the death of the affected insects. If applied to humans, chlorinated organic agrochemicals cause poisoning symptoms, such as headache, dizziness, loss of appetite, nausea, diarrhea, sweating, dyspnea, convulsion and hepatitis.

How were chlorinated organic agrochemicals, which have chlorine in their compositions, including tetrachloro-dioxin (TCDD), the most poisonous chemical compound that humankind has ever produced, and the abovementioned DDT and BHC, created? Chlorine is produced via the electrolysis of sodium chloride, or NaCl. We can obtain sodium hydroxide (NaOH), an important raw chemical material, by sodium chloride electrolysis, but this produces chlorine as a byproduct. Chlorine does not have any particular usefulness, whereas sodium hydroxide is indispensable in the modern chemical industry, as it can be used in construction material such as ferro-concrete or the manufacturing process of paper and soap. However, companies cannot abandon byproducts. The industrial circle had been struggling with the difficult problem of how to deal with chlorine gas as a byproduct, or how to put it to good use. They resolved this problem by combining hydrocarbon, produced in the gasoline production process, and chlorine gas. It was very convenient from the perspective of industrial use. In this way, large amounts of chlorinated hydrocarbon compounds were synthesized, and then used as agrochemicals.

In the case of chlorofluorocarbons and Freon gas, chlorine atoms contained in such gases led to the depletion of the ozone layer. Then, should we avoid all products containing chlorine? If we were to attempt to do so, we would need to reconsider sodium hydroxide production methods, or how to eliminate sodium chloride electrolysis. This would entail the development of a new way of sodium hydroxide production, without using sodium chloride. This is a challenge chemists are facing.

Pest control by attractants: An eco-friendly alternative

Driven by the need for new pest control methods to replace the mass spraying of organophosphorus or chlorinated organic agrochemicals, the development of eco-friendly and harmless alternative compounds,

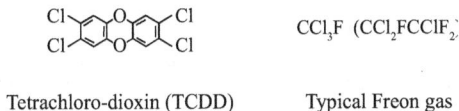

Tetrachloro-dioxin (TCDD)   Typical Freon gas

Figure 12.3: Structural formulae of TCDD and typical Freon gas

applied in small quantities, has begun. This alternative method utilizes insects' sex pheromones. It used to be thought that insects act according to their instincts. Recent discoveries, however, have revealed that their activities are all regulated by chemical substances. Assume that an ant accidentally found a pile of sugar. It is impossible for one ant to carry all the sugar back to its colony. This ant thus returns to the colony to seek help. On the way back, the ant releases so-called trail pheromones, a very simple compound, which has a pyrrole ring. This well-designed pheromone has moderate volatility, and trails disappear after a certain amount of time. If the pheromone trails remained after all the sugar were gone, the ants would travel back and forth for no reason. The workings of nature are well designed indeed.

Pheromones are defined as substances produced in vivo and discharged with very unique physiological effects on the same species, sometimes even on different species, even in a small amount. 'Hormone' is a similar term. Hormones are substances produced by special kinds of animal and plant cells, and have physiological effects on remote tissue cells. One of the typical hormones is insulin. Insulin is secreted by the pancreas and conveyed to the bloodstream, where it functions to control blood glucose levels. The human body secretes not only insulin, but also many kinds of hormones such as testosterone (male hormone) and estradiol (female hormone). Nevertheless, the human body does not produce any equivalent hormone that insects produce.

Returning to the topic of harmless alternative compounds to agrochemicals, here is an example based on insect sex pheromones: the sex attractant secreted by female silkworm moths. This attractant is a simple compound called bombykol, belonging to the higher alcohols and having two functional groups, namely olefin and hydroxyl. When this attractant is sprayed over a field, male silkworm moths confuse it with the presence of female moths. As

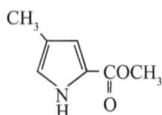

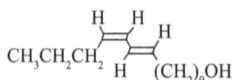

Ant's trail pheromone   Sex pheromone of female silkworm moths (Bombykol)

Figure 12.4: Structural formulae of trail pheromone of ants and Bombykol

a result, male moths are not able to mate with female moths. Based on this principle, we can exterminate insects harmful to the crops. Necessary amounts of such compound are very small. The exact figure on the use of bombykol is unknown, but in terms of the trail pheromone released by ants, just 0.3 mg is enough to mark a trail as long as the circumference of the Earth. In this way, pheromones are incomparably more effective than conventional agrochemicals, which need to be sprayed in large quantities. Henceforth, the development of these kinds of harmless and eco-friendly alternatives to agrochemicals is expected to become popular.

Pharmaceutical contribution to lifespan: Synthesis of chiral drugs

Because of the plague epidemic, which spread from China across Europe and Africa, the world population temporarily decreased in the fourteenth century. By contrast, the human population and average lifespan has increased dramatically since the 1800s. This may have been caused by the improvement in nutrition based on stabilized food production processes and by overcoming infectious disease through the development of vaccines. The invention and development of drugs able to cure diseases has significantly contributed to the population growth and the expansion of the average human lifespan. Here, I touch upon the synthesis of chiral drugs. A compound called thalidomide was responsible for incidents that occurred in Europe and Japan during the 1960s in relation to the harmful side effects of morning-sickness drugs.

Protein is an essential component for every creature on this planet. Almost all functions conducted by living cells require proteins. Protein is a filamentous polymer, consisting of twenty kinds of covalently bonded amino acids. All amino acids have the same basic

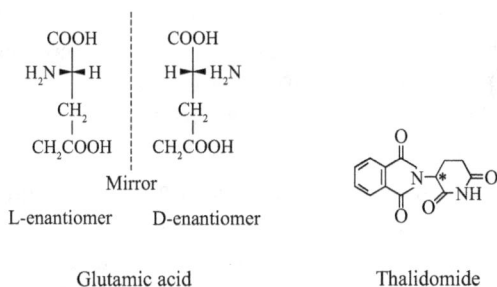

Figure 12.5: Structural formulae of glutamic acid and thalidomide

structure: the amino group, carboxyl group, hydrogen atom and R group (alkyl or aryl group) bond to a carbon atom in the center. A carbon atom that bonds to four different groups is called asymmetric carbon. A compound that has asymmetric carbons, such as glutamic acid (see Figure 12.5), has two enantiomers, namely D-enantiomer and L-enantiomer. L-glutamic acid is responsible for umami (savory taste), whereas D-glutamic acid is not. Proteins and peptides constituting our bodies are made of just one amino acid enantiomer. Think of this as the relationship between gloves and hands. Latex gloves can be worn irrespective of right or left. However, in the case of leather gloves, there is a distinction between right and left, and the left-hand glove does not fit the right hand. The reason why only one glutamic acid enantiomer is responsible for umami, as mentioned above, is that it fits in with taste buds on the tongue surface made from proteins, while the other enantiomer does not. The same applies to the relationship between drugs and the structure of disease-affected areas. Antibiotics and drugs cure diseases through an effect called enzyme inhibition, which immobilizes particular enzymes. If you think of antibiotics as leather gloves, and enzymes as the right or left hand, you'd understand that antibiotics need a certain asymmetric carbon structure. Before thalidomide related problems emerged, both of the enantiomers – corresponding right- and left-hand glove – were sold and taken altogether, even though only one enantiomer was effective. When synthesized in vitro, both right- and left-hand enantiomers were produced in the same proportion. They were sold together, because it was time-consuming and thus economically disadvantageous to separate them. The enantiomer ineffective in curing morning sickness was thought to be harmless, but in fact in the case of thalidomide it was potentially harmful: the non-effective enantiomer had teratogenicity, causing birth defects affecting the limbs. In the wake of such incidents, Japan's Ministry of Health no longer approves the use of medical drugs with asymmetric carbon unless they contain only the effective enantiomer.

In the previous paragraph, I wrote that right and left enantiomers will be produced in vitro in the same proportion, and it is thus impossible to synthesize only one kind. In 1827, Wöhler succeeded in converting ammonium cyanate, an inorganic compound, into urea, an organic compound. This finding went against the mainstream theory of the time, which stated that organic compounds could only be produced by living things. Likewise, it had been thought that only living organisms were able to produce one enantiomer without producing the other. Drs. Ryoji Noyori and Sharpless challenged this theory and succeeded in producing only one type of enantiomer. They received the Nobel Prize in Chemistry in 2001, along with Dr. Knowles. Their discovery has contributed significantly to the manufacturing of commercially available drugs. The research conducted by Dr. Negishi and Dr. Suzuki on 'palladium-catalyzed cross couplings in organic synthesis', which brought them the Nobel Prize in Chemistry in 2010, has also contributed to drug manufacturing to the extent that it is said to be unthinkable to synthesize drugs without this synthetic method. In this way, developments in organic synthetic chemistry have supported the extension of the average human lifespan by drug synthesis.

While drugs have been very useful in curing diseases, it takes an enormous amount of time and money to launch a new drug. It is said that it normally takes twenty years and twenty billion yen for a completely new compound (drug) to go through a research phase, clinical trials and eventually become available at hospitals for the treatment of patients. It is expected that the progress in iPS cell research will reduce the timeframe and costs of drug development. Yet, no matter how medical technology develops, and no matter how methods for drug synthesis are found, they will be of no use if we cannot secure enough energy (electricity or chemical materials) to implement them. Securing sufficient energy is the key to the survival of humanity.

Conclusion

Earth formed around 4.6 billion years ago. On the other hand, *Homo sapiens* appeared around 200,000 years ago. If we squeeze Earth's history into one calendar year, then the appearance of the human race would occur around 11:40pm on December 31. How can we make human history last? In other words, what can we do to maximize our existence? This is the overarching theme of Human Survivability Studies (HSS). What is survival? Although the meaning or interpretation of the purposes of survival might differ from person to person, I think we need a common understanding in order to build up the new scientific field of HSS. *D'où venons-nous? Que sommes-nous? Où allons-nous?* (See Chapter One).

How long human beings can exist depends on what kind of environment we live in. Shall we imagine a world with superabundant materials, even more so than now, or, a simple world like the primitive age, rewinding history? Also, in what timeframe should we consider our survival – a few hundred years from now, several thousand years or even tens of thousands of years? As our lifespan is only about one hundred years, it is extremely difficult for us to see into the distant future. As someone who studied organic chemistry in the current era, which future generations might call 'the plastic age' or 'the polymer age', I cannot imagine what will happen in a few thousand years, considering the fossil fuel reserves from which plastics are made. On the other hand, there is still room for growth in terms of agronomic performance, and we can anticipate that food production will be able to sustain the world population, even if it reaches ten billion. Increased food production has been made possible thus far, not due to the expansion of arable land, but because of the increase in crop yield per unit area. The output per unit area is not particularly high in Asia and Africa. It is possible to increase the yield, and thus secure enough food, by improving barren red clay. However, this presupposes a situation where climate change is not significant. If abnormal temperature rises or natural disasters like typhoons occur repeatedly, food production would decrease significantly, potentially leading to the recurrence of eschatology.

The above discussion on population and food problems reflects my vision of an ideal world in fifty to 100 years. As challenges about human survival become entangled, it is impossible to achieve an easy resolution. However agricultural technology, medical technology and relevant science develop, they are not sufficient to allow humankind to exist forever. Energy is essential in utilizing scientific technology, existent or novel. Eventually we have to face the energy problem, as will be discussed in Chapter Thirteen. Primary energy sources used worldwide today are fossil fuels, namely coal, oil and natural gas, and nuclear power generated from natural uranium. These are all natural, non-renewable materials. They will run out in the not too distant future, say 100–200 years from now. The world's energy consumption is expected to keep increasing at a more rapid pace than population growth. Based on such predictions, someone has to discover a fifth primary energy, put it to practical use, and bring about innovation, which will enable us to have a steady supply of such energy. Currently, there are vigorous efforts underway in research on wind power generation, geothermal power generation, nuclear fusion, biofuel and solar power generation. I must say, however, that the road to the solution is still long, because all of these new energy sources face high technological and financial hurdles. The fifth energy source must be found before the depletion of the four current sources, and thus we, the current generation, bear the responsibility of conserving the current energy sources as much as possible, in order to make them last.

In the twentieth century, significant effort has been devoted to the development of scientific technology and the discovery of new technologies, as it was believed that science would bring happiness to humankind. Is the world today, filled with materials made possible by such advancements, a happy environment for humankind? What about the twenty-first century? Is a life rich with chemical materials rich in a mental sense? Can each individual be happy in such an environment? It is uncertain how long human history will last, but I hope it will be happy as long as it continues. We must have a discussion on what constitutes a 'happy environment', beyond the borders of the study of arts and sciences.

13 The Origin and Nature of Resources and Energy

Shigeki Sakurai

The Earth continuously receives solar radiation from the sun and releases radiant heat back into outer space. Humans make use of these energy flows for the development of society and its economy in the form of electricity, light, heat and so on. These ultimately turn into heat energy that flows into low-temperature regions, resulting in overall homogenization.

The answer to the question on how humans should utilize such energy for future developments revolves around the ways to improve and optimize such energy flows (hereafter, 'generating energy as a flow').

For example, renewable energy such as solar photovoltaic power generation utilizes light, which is a high-quality form of energy emitted by the Sun. Wind and hydraulic power generation also utilize similar mechanisms. Biomass energy resources are generated by utilizing light energy from the Sun and CO_2 fixation. Conversely, geothermal energy originates from the Earth itself in the form of magmatic heat.

Fossil resources and energy such as petroleum, natural gas and coal are also generated from the Earth's ecosystem. In other words, solar energy fixated by different organism groups is turned into highly beneficial and high-quality resources and energy under the high temperatures and pressure deep underground for long periods ranging from tens to hundreds of millions of years. Such resources as well as biomass resources can be used not only as fuels but also as raw material resources (conversion to chemical raw materials and the like; noble use of oil) after being processed by human beings.

Mineral resources such as copper and rare earth metals have been found in the deep and wider areas below the Earth's crust. Since the formation of the Earth, these elements have gradually become enriched in certain locations in the Earth's crust and are now available in the form of metallic ore deposits. These deposits are extracted to the surface and refined into metals and subsequently processed for use in the manufacture of materials and products (railways, ships, automobiles, home electronics appliances and batteries) essential for human use. In addition, these goods are recyclable and reusable. Hence, their efficient and effective utilization is of utmost importance (hereafter, 'manufacturing materials as a cycle').

What, then, is the origin of nuclear energy resources? Nuclear energy is neither solar nor magma-derived energy. It is extracted as significantly high-density energy with changes in atomic numbers that occur at the time of energy generation, i.e., after the transformation of atoms into other atoms (nuclear reactions). It should be noted that the use of nuclear energy is fundamentally different in nature from the use of fossil fuel resources, which are burned (chemical reaction) to generate thermal energy. It is also different from the extraction of useful minerals from mines and their refinement into metals.

Another energy source in the nuclear energy category is nuclear fusion, which is also accompanied by atomic number changes. In nuclear fusion, energy is introduced in the reaction process from the outside, and the changes in atomic numbers generate high-density energy. The energy generation reaction mentioned in the previous paragraph is accompanied by an atomic number decrease, and the energy generation reaction with nuclear fusion is accompanied by an atomic number increase (Akimoto 1995).

Therefore, discovering ways to secure these resources and energy with the aim of sustainably developing humanity in the future is undoubtedly one of the most important global issues (hereafter, 'securing resources as a foundation'). It is essential to address these global issues on the basis of our understanding of the origin and nature of each resource and form of energy as mentioned above. It is also necessary to keep in mind that the extraction of fossil fuels and mineral resources causes a burden on the global environment and generates greenhouse gasses at the stages of production, processing and use. Thus, it is indispensable to establish a society that can comprehensively address these issues by means of both soft and hard technologies, including international policies and measures and institutional design, and advanced, collaborative and technological development. In other words, the key point is how to orient and establish a society that makes it possible to enhance

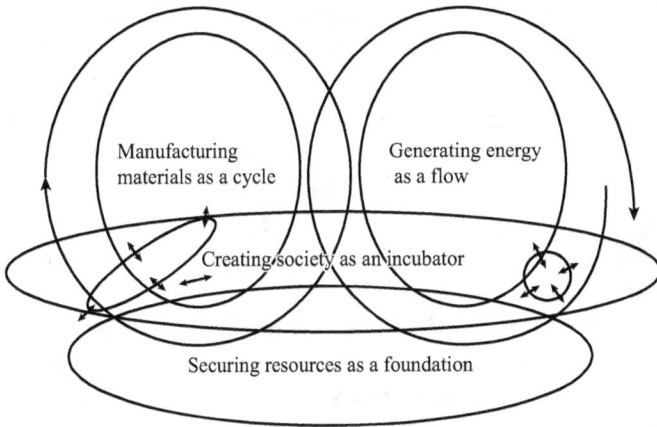

*Figure 13.1: Four elements of survivability for coping with global-
scale resource and energy issues*

human survivability in the face of global-scale resource and energy
issues (hereafter, 'creating society as an incubator').

Survivability (Human Survivability Studies) in the face of global-
scale resource and energy issues discussed here refers to the global
ability to realize 'securing resources as a foundation', 'generating
energy as a flow', 'manufacturing materials as a cycle' and 'creating
society as an incubator', thereby pursuing optimal performance (see
Figure 13.1).

Global-scale resource and energy issues

In this section, in order to understand the current global situation of
resource and energy issues, we summarize the major elements, i.e.,
the supply of and demand for resources and energy and international
efforts addressing global warming.

Resources and energy supply and demand

According to the International Energy Agency's (IEA) energy
supply and demand outlook, the global energy demand anticipated
in 2035 is 16.9 billion tons (converted to petroleum tonnage), which
corresponds to about 1.3 times of the 12.71 billion tons recorded in

2011. Of the 4.19 billion-ton increase in demand (16.9 minus 12.71 billion tons), that from non-OECD countries is estimated to account for about 96% (4.03 billion tons) of the total demand. In 2035, China's energy demand is expected to reach 24% of the gross global demand (about 1.5 times its demand in 2011), and India's is expected to reach 9% of the gross global demand (about 2.1 times its demand in 2011). Thus, brisk energy demand is expected in the future, particularly from developing countries in the context of their rapid economic growth (IEA 2013).

Next, we summarize the outlook for fossil fuels such as petroleum, natural gas and coal as well as atomic energy fuel sources by changing the focus to the supply side in terms of medium- and long-term resource depletion.

The recoverable reserves of petroleum (excluding oil sand) were 1.6689 trillion barrels as of the end of 2012. About half of these reserves exist in the Middle East, and about one-third are in Venezuela and the Americas, such as Canada. The minable years for these reserves are estimated at fifty-three years. In addition, the future development and production trends of shale oil, an unconventional oil resource, also require close attention.

As of the end of 2012, the recoverable reserves of natural gas were 187 trillion m^3, of which more than 40% exist in the Middle East, about 30% in Europe and Russia and about 30% in other former Soviet republics. The minable years for these resources are estimated at fifty-six years. Notably, in the US, the production of shale gas, which is an unconventional resource, is expanding rapidly, with many LNG projects being scheduled. Moreover, the promotion of coalbed methane (CBM) production is also anticipated.

According to sources such as BP's 2013 statistics, recoverable reserves of coal amount to 860.9 billion tons, of which the US accounts for about 28%, Russia about 18% and China about 13%. Coal is widely present around the world, with its minable years being estimated at 109 years.

According to the joint report published by the OECD, Nuclear Energy Agency (NEA) and International Atomic Energy Agency (IAEA) titled 'Uranium 2011: Resources, Production and Demand' for example, the recoverable reserves of uranium are 7.0966 million tons, of which Australia accounts for 24.5%, Kazakhstan 11.6%, Russia 9.2% and Canada 8.7%. This resource is distributed worldwide, and the minable years are estimated at 122 years (Ministry of Economy, Trade and Industry 2014).

According to the above supply-demand outlook, a tight situation for the noted fuels in the medium- and long-term can be easily assumed.

Regarding nuclear energy, the accident at the Fukushima Daiichi Nuclear Power Station triggered by the Great East Japan Earthquake on March 11, 2011 has significantly affected Japan's nuclear energy policy and even the policies of some European countries. Therefore, efforts for reviving and reconstructing Fukushima Prefecture, including decommissioning and countermeasures against contaminated water, should be the utmost priority among nuclear power generation issues in Japan. At the same time, there are many other issues that should be deliberated, such as the nuclear fuel cycle (including radioactive waste disposal) and the resumption of operations at other nuclear power plants. In other words, these issues are part of the overriding tasks that should be discussed nationwide involving the general public as work that constitutes the core of establishing Japan's future energy policies. The global nuclear energy trends, as noted above, and Japan's mission as a member of the international community should also be adequately taken into consideration (Cabinet Decision 2014).

Fundamental outlook in terms of global environmental issues

As discussed above, the medium- and long-term demands for global resources and energy are expected to grow steadily, while the recoverable reserves of conventional fossil fuels are not sufficient to meet this demand. In this context, to meet the medium- and long-term demand for primary energy, it will be essential to expand and promote the current introduction of renewable energy and use of nuclear energy as well as to promote the development and introduction of unconventional fossil fuels.

However, the end of the twentieth century saw a series of controversies originating with the *Limits to Growth* published by the Club of Rome in 1972 and the report issued by the Brundtland Commission in 1987. Following the Earth Summit in 1992, global environmental protection and sustainable development came to the fore as two of the most important global tasks, which led to the establishment of the Kyoto Protocol in 2005 (Meadows et al. 1972; World Commission on Environment and Development 1987).

Regarding efforts towards addressing the pressing issue of global warming, various deliberations have occurred on measures against future climate change, such as the 19th Conference of the Parties to the United Nations Framework Convention on Climate Change (COP 19) of November 2013, which brought about the post-2020 framework (post-Kyoto Protocol framework). One of the key points is the reduction of greenhouse gasses. However, it would be difficult for policy efforts alone to achieve a global reduction in greenhouse gas emissions by half by 2050 and reduce greenhouse gas concentration in the atmosphere to CO_2-converted 450ppm by the end of the twenty-first century. Medium- and long-term economic growth is expected in newly emerging nations such as China and India as well as other developing countries in Asia and Africa. Hence, it will be impossible to offset greenhouse gas emissions solely by means of promoting and enhancing energy conservation, highly efficient and clean utilization of fossil fuels, further introduction of renewable energy and the expansion of atomic energy utilization. This was noted in the Fifth Assessment Report and Third Working Group Report of the Intergovernmental Panel on Climate Change (IPCC) in April 2014 and by the IEA/ETP (Energy Technology Perspective) in May 2014. Significantly, there has been a new emphasis on the need for a considerable introduction of carbon capture and storage (CCS) measures, such as fossil energy utilization combined with CCS and bio-energy combined with CCS (BECCS) (IPCC 2014; IEA 2014).

Global challenges for resource and energy issues

Keeping in mind the origin and nature of resources and energy as mentioned above, we now discuss Japan's current global position in addressing global-scale resource and energy issues. Several specific projects are also referenced in terms of survivability (Human Survivability Studies) and global challenges.

Basic point of view: 'On the same boat'

There is an uneven regional distribution of various resources such as fossil fuels and metallic minerals. The potential quantities of some of these resources are small, while resources are not necessarily extracted in the countries that consume them. However, it is undeniable that CO_2 is emitted during the extraction, refinement, processing and consumption stages, and that this causes global environmental problems. Moreover, the suitability of renewable energy such as photovoltaic, wind and hydraulic power differs according to geographical features as well as weather and climate. Every country, regardless of its development status, is endeavoring to secure necessary resources and energy to maintain its production activities and social system, and to achieve economic growth for further social development while paying attention to environmental problems. On the other hand, countries cannot avoid mutual impacts as multinational exchanges of goods, services, finance, information and so on are expanding in quantity and international rules are intensifying and becoming more complicated.

'We are on the same boat'

As stated in Chapter Two, the most important fact we human beings must be aware of is that we are all riding on the same vessel called Spaceship Earth. Nations have no choice but to pursue co-existence and co-prosperity without clinging to competitiveness, and they cannot get off this Spaceship Earth. In other words, the survivability of this vessel is our most important task.

Efforts toward survivability related to resources and energy include those made in response to the first oil crisis, such as the establishment of the IEA, the maintenance of petroleum reserves in cooperation with related nations, the prohibition on building new oil thermal power plants and the emphasis on the development of alternative-energy technologies to reduce reliance on petroleum. The Earth Summit in 1992 highlighted the concept of 'sustainable development' and led to the establishment of the IPCC and the framework for international cooperation on global environmental issues. The IAEA was founded to bring about the safe utilization of nuclear energy and nuclear nonproliferation as well as to provide a strict nuclear management mechanism.

In this way, resource and energy-related global coordination, cooperation and collaboration are indispensable. The ultimate objective is to maximize value creation in global-scale resource and energy supply-demand chains as well as to minimize environmental burdens.

For this objective, nations are working together in various ways as discussed in Part V of this volume (Chapters Two and Three). In particular, to address resource and energy issues, it is essential to promote diverse efforts and international collaboration ranging from the OECD/IEA, IPCC/COP and IAEA frameworks to bilateral and multilateral cooperation, treaties, economic and trade agreements, technology transfer, development cooperation, R&D cooperation, education and human resource cultivation.

Japan's position

In the current situation, the demand for fossil energy sources such as petroleum, natural gas and coal is anticipated to expand in the medium- and long-term. Hence, resource and energy issues including the diversification of energy sources, such as the introduction and promotion of renewable energy and the expansion of peaceful uses of nuclear energy, and the handling of global environmental issues are expanding, deepening and becoming increasingly complex as global issues. In this situation, Japan needs to secure its required resources and energy and at the same time is tasked with the international responsibility of appropriately coping with global environmental issues. However, as mentioned above, Japan has a significantly low resource and energy self-sufficiency ratio of 6.0%, due to its fragile resource and energy supply structure. Moreover, there is a need to reduce greenhouse gasses by further increasing the sophistication of Japan's energy-saving society, which is already at the top level internationally, and address the impacts of the Great East Japan Earthquake. Clearly, Japan is facing a very serious situation (Ministry of Economy, Trade and Industry 2014).

In the medium- and long-term, the current situation in Japan is one that other nations cannot avoid, regardless of whether they are in a developed, newly emerging or developing state. It is anticipated that this situation will become more serious in the future. We believe that taking on the challenge presented by these issues by actively playing a leading role for the sake of Japan as well as other nations is a responsibility and position that this country should assume as a crew member of Spaceship Earth.

To make its position credible, and particularly because it is a country poor in resources and energy, there is a strong expectation that Japan will use its globally top-level technologies to gain a foothold in the development of exceptional energy saving and low-carbon technologies in different fields including power generation, steel, automobiles, home electronic appliances and so on. Japan is also expected to take necessary action, including: the promotion of renewable energy demonstration projects for offshore wind power generation and advanced large storage batteries; the promotion of the 3Rs (recycle, reuse and reduce) and the development of rare earth materials and the like that can act as substitutes for rare metals by making use of technologies for manufacturing from resources and materials; the establishment of energy management systems such as BEMS, HEMS, CEMS and demand response; the promotion of smart city project demonstrations; the dissemination of stationary fuel cells (Ene Farm); and the establishment of a hydrogen-based society, such as by commercializing fuel cell electric vehicles. In this way, Japan will greatly contribute to global technology transfer, as well as actively cultivate global markets.

For Japan to live up to its responsibility in the future, ensuring the effectiveness of survivability (Human Survivability Studies) is deemed to be essential. In the next section, we will discuss some specific proposals for its global challenges.

Japan's global challenges

From 'smart mining' and 'sustainable development' to 'survivability'
With regard to resource procurement as the foundation, as mentioned above, securing resources simply means extracting them from underground, no matter how the most advanced technologies are used.

Currently, as a typical case of mineral resource development, the major focus is on the mining of resources without damaging the ecosystem and without emitting environmental pollutants to the greatest extent possible (without damaging the Earth) under the concept of sustainability. Moreover, when Japan provides development support to a developing country, it gives wide-ranging and comprehensive support including human resource cultivation, mineral marketing support and resource processing technology transfer within the framework of ODA and the like (smart mining and sustainable development; Sakurai 2010).

What, then, is envisioned or should be envisioned when further considering such development support in terms of survivability, i.e., in terms of the development target areas and their future outlook?

For example, in the case of an open pit mine (strip-mining site), particularly one that is a few km square and more than 100 m deep, it would be favorable to develop the site as an artificial lake and turn the surrounding area into a natural park, rather than simply restoring the land to its original state using landfill, planting and the like. Moreover, it would be wonderful if an area from which mineral resources have been extracted could, after several decades, be turned into a resort area able to accommodate people from large cities and provide vitality and energy. However, the important point is surely not to plan such redevelopment once the mining is complete, but to envisage medium- and long-term intergenerational sustainability (active involvement of the next generation) at the initial mine development phase.

In any case, it is an undeniable fact that mining resources and their utilization are indispensable global activities for human beings.

'The marriage of biomass and coal'

The above phrase may sound very strange to some readers. Conventionally, the co-combustion of biomass during coal-fired power generation has been widespread in Japan, particularly since the Renewables Portfolio Standard (RPS) law took effect. This was originally intended to reduce CO_2 emissions. As a 180-degree change in the ordinary way of thinking, here we refer to the Advanced Biomass-Coal Co-gasification Project (called the ABC Project), based on the idea of promoting the utilization of unutilized biomass resources. 'Biomass instability' refers to that caused by an inadequate amount of collected biomass that has low calorific value due to high water content and other factors. The fundamental vision of this project is to appropriately complement this biomass instability by means of coal, thereby promoting and expanding the introduction of biomass and coal utilization. As explained above, in the process of making coal-fired power generation more sophisticated, gasification technology has been practically established, and because coal and biomass are both solid raw materials, the synergetic effects of their co-gasification are expected to be validated technologically. Moreover, if CO_2 generated in the process of co-gasification is captured and stored

(CCS), 'carbon minus' is achievable. If this is realized, projects can be anticipated that use not only wood waste but also raw sewage, sewage sludge cakes and so on. In other words, co-gasification can extract high-quality resources and energy such as H_2 and CO, which also provides the possibility of contributing to chemical raw materials, fuel cell electric vehicles and a hydro society. Projects in developing countries with abundant biomass and coal resources are expected to play a leading role in the promotion and sophisticated utilization of biomass resources and international cooperation projects in such coal-producing countries.

For example, such efforts are ongoing, beginning in FY2012, as the Tri-Generation System Demonstration Project Utilizing Biomass Produced from Snow-Falling Mountainous Areas. This is being carried out with the Japan Coal Energy Center and the Yokote Shinrin Kumiai (Yokote City Forestry Association) playing leading roles, and with the cooperation of Akita Prefecture, Tohoku Electric Power Co., Inc. and so on (Japan Coal Energy Center 2014).

Advanced clean coal cycle: 'Ultimate coal-fired power generation'
According to the IEA's WEO-2013, the share of demand for coal-fired power generation is expected to remain the largest in the global demand for primary energy in 2035.

With regard to the global performance of Japan's coal-fired power generation, its power generation efficiency is at the highest level. It has been estimated that if the coal-fired power generation plants of major nations such as China and India were replaced by those with the same efficiency level as those in Japan, it would achieve annual CO_2 emissions reductions amounting to the total of all emissions from coal-fired power generation plants in Japan (Agency for Natural Resources and Energy, 2014). In addition, Japan is also at the top level in the reduction of NOx and SOx emissions.

In this context, it is particularly worth mentioning the Integrated Coal Gasification Fuel Cell Combined Cycle (IGFC), which is an advanced coal-fired triple power generation system. In this system, coal is gasified and triple power generation is performed using gas turbines, steam turbines and fuel batteries, resulting in increased power generation efficiency by more than 55%. Moreover, CO_2 separated during the gasification process is captured and stored (CCS) (Japan Coal Energy Center 2013). If this system is realized in the future, CO_2-free coal-fired power generation at the world's highest level of efficiency will become possible. By deploying this system in major coal-fired power generation countries such as China and India, it will be possible to achieve remarkable CO_2 reductions around the world. The significance of the steady commercialization of this system is also very high in raising the effectiveness of the CCS scenarios described in the Fifth Assessment Report of the IPCC and the IEA's Energy Technology Perspective 2014. The practical application of this system is a global challenge Japan can contribute to.

International expansion of 'survivable communities'

Discussions regarding resilience have been particularly active since the Great East Japan Earthquake. Especially in Japan, the fragility of large-scale energy systems has been frequently discussed since the earthquake, and a distributed energy system is gaining attention in terms of resilience. In this context, the verification and establishment of so-called 'smart communities' are being promoted in a certain scale of community with the participation of various players such as energy suppliers, companies and end users (such as household electricity users). Smart communities are based on energy management systems such as those combining renewable energy and storage batteries, stationary fuel cell power generation systems, cogeneration systems and BEMS/HEMS (Ministry of Economy, Trade and Industry 2014).

If such smart communities are developed in emerging and developing countries that produce resources and energy, it is expected that certain communities in such countries would be able to utilize renewable energy including biomass, geothermal and solar energy, and also produce energy resources including natural gas, coal, CBM and shale gas at the same time. If Japan can lead other countries in introducing advanced, integrated projects in which renewable energy and locally-produced fossil fuel resources are incorporated, it can globally establish and expand the use of smart, resilient communities that follow local production for local consumption type energy use (i.e., viable communities) that are uniquely achievable in such countries.

Energy R&D challenge for collectively pioneering the future

Following the discussion regarding the medium- and long-term outlook for petroleum, natural gas, coal, renewable energy and nuclear energy, looking ahead into the future the next subject is advanced R&D for experimental thermonuclear reactors (100 to 200 years from now).

Currently, an international nuclear fusion energy project called the International Thermonuclear Experimental Reactor (ITER) project, whose members include Japan, the European Union, the US, Russia, China, the Republic of Korea and India, is endeavoring to realize a thermonuclear experimental reactor for the future of humanity by bringing together the knowledge of each country. The objective of ITER is to realize nuclear fusion energy by verifying its scientific and technical potential on the basis of international agreements and through the construction and operation of the ITER experimental nuclear fusion reactor. Verification of 500,000 kW-class heat generation is scheduled for 2020, with the target of commercialization in 2050 (Ministry of Education, Culture, Sports, Science and Technology, 2014). Fuel resources for the reactor are naturally present heavy hydrogen (deuterium) and lithium, estimated to be available for about 1.5 million years when converted to the amount of gross power generation worldwide. In practical terms, there are various challenging engineering hurdles to be cleared before commercialization of the project (Japan Society of Plasma Science and Nuclear Fusion Research 2007).

Finally, how should humanity deal with this ultra-large-scale international project for realizing an experimental nuclear fusion reactor, the first such attempt by human beings, which is the ultimate goal for advanced energy R&D? We believe that concerted efforts for the project will lead to global survivability for securing resources and energy for the future of humanity, and Japan's strong promotion of the project is in itself a global challenge. Readers, what are your thoughts on this?

Incidentally, the word '*iter*' means 'the way' in Latin, symbolizing hope for a path that leads to the realization of nuclear fusion and international cooperation for the sake of the Earth.

Part IV
Human Survivability Studies in Practice

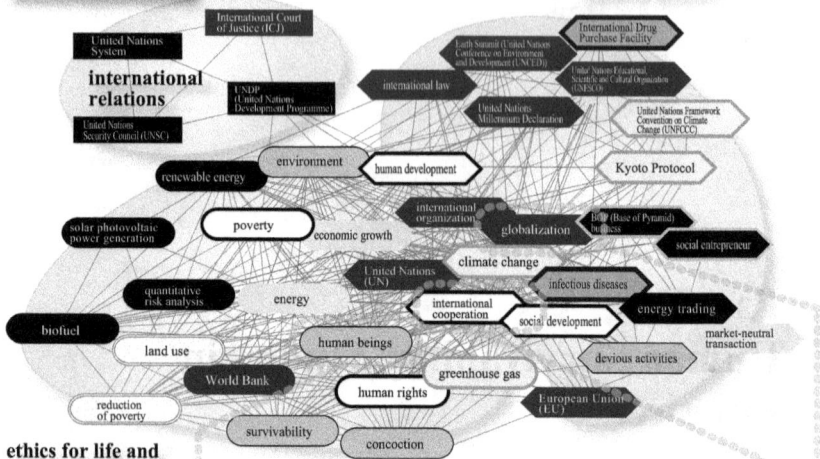

global issues

climate change and
natural disaster

international
relations

International Court
of Justice (ICJ)

United Nations
System

International Drug
Purchase Facility

Earth Summit (United Nations
Conference on Environment
and Development (UNCED))

United Nations Educational,
Scientific and Cultural Organization
(UNESCO)

UNDP
(United Nations
Development Programme)

international law

United Nations
Millennium Declaration

United Nations Framework
Convention on Climate
Change (UNFCCC)

United Nations
Security Council (UNSC)

environment

human development

Kyoto Protocol

renewable energy

international
organization

BOP (Base of Pyramid)
business

solar photovoltaic
power generation

poverty

economic growth

globalization

social entrepreneur

quantitative
risk analysis

energy

climate change

United Nations
(UN)

international
cooperation

infectious diseases

social development

energy trading

biofuel

land use

human beings

greenhouse gas

devious activities

market-neutral
transaction

World Bank

human rights

European Union
(EU)

reduction
of poverty

survivability

concoction

ethics for life and
the environment

Chapter 16

Chapter 15

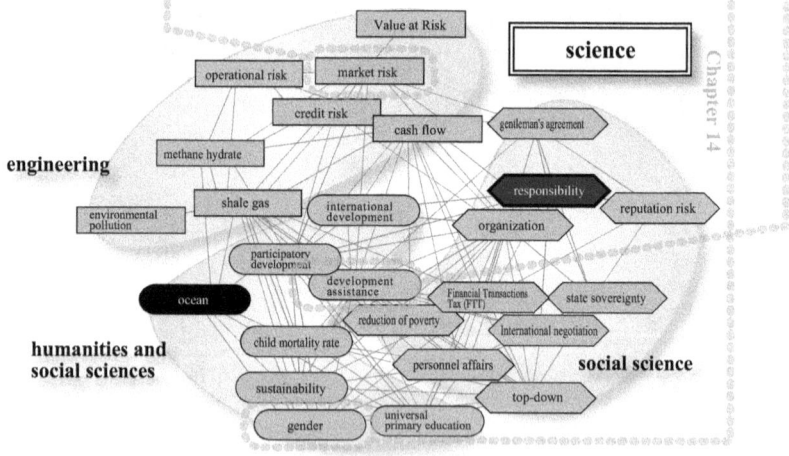

Value at Risk

science

operational risk

market risk

credit risk

cash flow

gentleman's agreement

engineering

methane hydrate

shale gas

international
development

responsibility

reputation risk

environmental
pollution

organization

participatory
development

development
assistance

Financial Transactions
Tax (FTT)

state sovereignty

ocean

reduction of poverty

International negotiation

humanities and
social sciences

child mortality rate

personnel affairs

social science

sustainability

top-down

gender

universal
primary education

Chapter 14

Introduction to Part IV

Chapter Fourteen discusses the definition of and trends in international cooperation. Additionally, this chapter introduces actors in this field – new actors, such as local governments and private companies, as well as conventional actors like the central government and JICA (Japan International Cooperation Agency). The last part of this chapter talks about the international cooperation facilitated by GSAIS students through their overseas internships.

Chapter Fifteen deals with the roles played by international organizations in global society. The first part of the chapter explains the definition and existence of international organizations, and highlights the current challenges they face. The second part takes a general view of the partnership between international organizations and civil society, companies and universities. Lastly, Chapter Sixteen discusses one of the most important issues facing the modern world, namely the problem of risk and its management in socioeconomic activities. As specific examples, the chapter examines the problems of risk and its management in corporate activities, resource development and scientific/engineering ethics. It points out that obtaining solutions from the existent and compartmentalized academic system is difficult, and concludes that Human Survivability Studies has the capacity to resolve this issue.

14 International Development: Development Assistance and Self Reliance

Senichi Kimura and Michiyo Hashiguchi

Introduction

Human beings and global society face various threats and crises. It is foreseeable that the situation will become more severe in the near future. As the migration of people and movement of things, capital and information increases on a global scale, the nature of threats such as climate change, natural disasters, conflicts and terrorism will diversify. As a result, we, wherever we live, increasingly realize threats directly and/or indirectly even in everyday life.

In the field of international cooperation, Japan and other developed countries are striving to improve the lives of people in developing countries through investing in human, material and financial resources. International cooperation in the form of knowledge and technology transfer from developed countries to developing countries is arguably necessary and will continue in the future.

However, due to the increasing complexity of the issues faced by humanity, we now have to rethink international cooperation. Transferring knowledge and technology from developed to developing countries might not be sufficient. In relation to threats and crises, there is no distinction between developed and developing countries. We all share responsibility because we are involved in causing threats and are affected by them. We are all in the same boat.

There is not a single solution to overcome the vast issues we face as a species. Every effort towards finding solutions should be made through trial and error. This is not an easy task to do. Interaction among people and mutual learning are becoming increasingly important. To do this we have to work with people of different cultures, habits and languages.

International development

In Japan, the term 'international cooperation' is often used to describe cooperation or assistance that crosses national borders. However, the term is not clearly defined. Some may think of international cooperation as that between/among governments. Others may think of the international activities of NPOs and NGOs.

In Japan, 'international cooperation' usually means assistance provided by developed countries to developing countries. JICA, an agency that implements the Government of Japan's ODA (Official Development Assistance), stands for 'Japan International Cooperation Agency'. JICA carries out assistance programs in developing countries. Meanwhile, in western countries terms such as 'development cooperation' or 'international development' are more familiar than 'international cooperation'. USAID, the agency that implements the US's ODA, stands for 'United States Agency for International Development', and DFID, that of the UK, stands for 'Department for International Development'.

When we discuss 'international cooperation', we need to understand the concept of 'development', which changes as time and society change. In the past, 'development' mainly meant economic growth. Development assistance programs focused on transferring financial resources and technology from developed to developing countries in order to promote economic growth.

Some developing countries have achieved economic growth, but at the same time this has resulted in distortions such as wealth disparity and widening inequity. As a result, after the 1980s, social and human development has been highlighted in addition to economic forms of assistance.

While some issues have been resolved or improved by international development activities, new types of threats such as disasters caused by climate change, infectious diseases like Ebola hemorrhagic fever, conflicts and terrorism have emerged. Under such circumstances, it is of utmost importance to identify new approaches to development to both sustain and facilitate the survival of humanity.

New trends in international development

Discussions on social problems began in the 1980s due to distortions caused by economic-oriented development. As a result, the importance of social development has been highlighted and it is recognized that civil society has an important role to play in development in addition to traditional players like governments and enterprises.

In the 1990s, the importance of human development was also emphasized as social development, as the situation surrounding poverty, conflicts and environmental problems became worse.

Human development and human security

The concept of 'human development' appeared in the UNDP's 'Human Development Report 1990' and 'Human Security' in 1994. It is defined as the process of expanding people's life choices and of strengthening capacity through the improvement of health, knowledge and skills, and the utilization of such skills. Through the human development process, the environment in which people can develop their potential and the constructive and creative life based on their needs and interests regardless of their income can be generated.

The concept of 'human security' is beyond that of conventional 'national security'. After the Cold War, domestic conflicts intensified worldwide, as did globalization, entailing flows of people, goods, capital and information across borders. Under such circumstances, 'human security' is needed in response to the complexity of both old and new security threats – from chronic and persistent poverty to ethnic violence, human trafficking, climate change, health pandemics, international terrorism and sudden economic and fiscal downturns. Such threats tend to acquire transnational dimensions and move beyond traditional notions of security that focus solely on external military aggression.

Secondly, 'human security' is required as a comprehensive approach that utilizes the wide range of new opportunities to tackle such threats in an integrated manner. Human security threats cannot be tackled through conventional mechanisms alone. Instead, they require a new consensus that acknowledges the linkages and interdependencies between development, human rights and national security.

Human security brings together the 'human elements' of security, rights and development. As such, it is an interdisciplinary concept that displays the following characteristics: it is people-centered, multi-sectoral, comprehensive, context-specific and prevention-oriented.

Participatory approach

Along with introducing the concepts of human development and human security, realizing them necessarily involves stakeholder participation in development. The idea of participatory development was first declared in the policy statement 'Development Assistance in the 1990s' by the Development Assistance Committee (DAC) of the Organization for Economic Co-operation and Development (OECD). This statement emphasized that the greatest number of people possible can be involved in decision-making regarding development and enjoy its benefits.

The methodology to promote participatory development has been arrived at through trial and error in the field, such as Participatory Rural Appraisal (PRA) and Participatory Learning and Action (PLA, etc.). It is, however, necessary to always consider the appropriateness of the participatory approach, whether it is superficial or not, and what its purpose is, because the decision-making process can be difficult and complex.

Governance

Although the term 'governance' has wide and distinct meanings, it is determined in the field of international development as a prerequisite for the realization of development, an element related to the effects and efficiency of development or a value and situation to be achieved through development assistance. Good governance is given importance as a precondition for the realization of development.

Governance has two aspects. One is as the basis of nations, such as the government's legitimacy, responsibility to the people, the guarantee of basic human rights and so on. Another aspect is governance as function, organization/institution, a capacity and a mechanism of the government that acts as a foundation to promote the participatory approach.

MDGs and SDGs

The Millennium Summit was held in September 2000 at United Nations Headquarters in New York. One hundred and forty-nine heads of state and government and high-ranking officials from over forty other countries attended. The main document, unanimously adopted, was the Millennium Declaration, which contained a statement of values, principles and objectives for the international agenda for the twenty-first century. It also set deadlines for many collective actions.

The world leaders who gathered at the Summit committed their nations to a new global partnership to reduce extreme poverty, and set out a series of time-bound targets with a deadline of 2015, known as the Millennium Development Goals.

The Millennium Development Goals focus on eight major issues: (1) eradicating extreme poverty and hunger; (2) achieving universal primary education; (3) promoting gender equality and empowering women; (4) reducing child mortality; (5) improving maternal health; (6) combating HIV/AIDS, malaria and other diseases; (7) ensuring environmental sustainability; and (8) developing a global partnership for development.

World leaders, participants and numerous stakeholders have worked towards achieving the MDGs by 2015. It can be said that some targets have been achieved while others have not. Necessary actions for remaining and emergent issues have been discussed as goals beyond 2015.

At the United Nations Sustainable Development Summit on September 25, 2015, the target year of the MDGs, more than 150 world leaders adopted the new 2030 Agenda for Sustainable Development, including the Sustainable Development Goals (SDGs).

The seventeen new Sustainable Development Goals, also known as the Global Goals, aim to end poverty, hunger and inequality, take action on climate change and the environment, improve access to health and education, build strong institutions and partnerships and more.

The SDGs have more ambitious agendas, seeking to eliminate rather than reduce poverty, and include more demanding targets on health, education and gender equality. They are universal, applying to all countries and all people. The agenda also includes issues that were not in the MDGs such as climate change, sustainable consumption, innovation and the importance of peace and justice for all.

International development actors

International development has been primarily carried out by implementing agencies such as JICA, international organizations and NGOs. In recent years, it has been expected that new actors for international development such as civil society organizations and private firms will play vital roles, because relevant fields and issues have become diversified and globalization has expanded.

Local governments in Japan do not function as international development actors because they are administrations dealing with domestic matters. They do, however, possess plenty of technical and administrative know-how, experience and skills in various fields such as water supply, sanitation, waste management, pollution control, public and reproductive health, social welfare, agricultural extension, primary and secondary education, vocational training, public transportation and so on for the provision of services to citizens in their local area. These capacities must be utilized in the development of developing countries. The system of Japanese local government itself could be a good example for developing countries where decentralization is in progress.

Private firms in developed countries have thus far participated in ODA projects via contracts ordered by governments or donor agencies. Recently, non-profitable activities by private firms as CSR (Corporate Social Responsibility) are increasing in developing countries. In addition, BOP (Base of Pyramid) business is also increasing as one of the new business models. In BOP business, a new significant potential market of four billion people exist, who form the bases of the economic pyramid and are potential consumers who can enjoy such new businesses established by private firms. In this new business model, it is possible for social entrepreneurs committed to social development to carry out international development activities without assistance or donations.

Universities are represented as the node of knowledge with ample and varied experience, knowledge, expertise and human resources. Universities in Japan are one of the vital actors of and have been playing important roles in international development. These include acting as the implementing body for technical cooperation projects (including SATREPS: Science and Technology Research Partnership for Sustainable Development), dispatching experts to developing countries and accepting foreign students and trainees and so on. They are also pioneering new approaches towards globalization, such as starting new courses etc. to educate students to work in rapidly growing developing countries. Hence, universities in Japan collaborate with JICA and other international development organizations in terms of both direct contribution

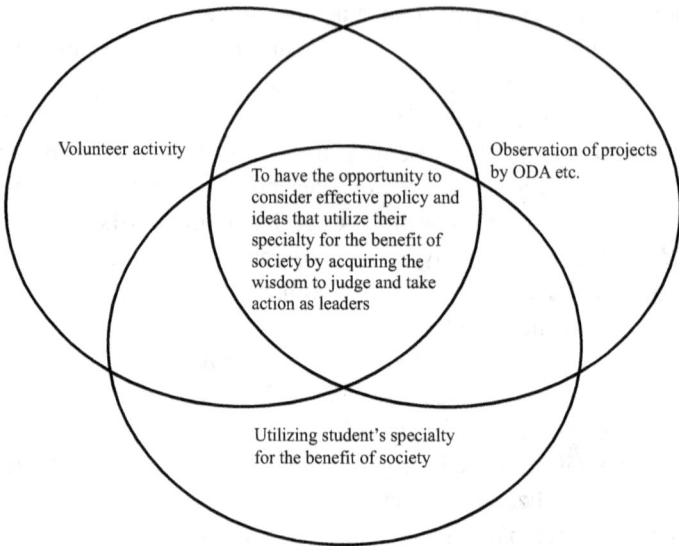

Figure 14.1: Concept of Overseas Internship

to development and the education of potential human resources for development.

GSAIS Overseas Internship Program

GSAIS (Graduate School of Advanced Integrated Studies in Human Survivability, Kyoto University) provides its second year students with an overseas internship program (OIP) conducted over summer as a service learning program. Through this program, students are trained to be global leaders.

In the service learning program students work as volunteers to be trained not only to learn the mindset, but also to understand various values, viewpoints, culture and customs as a leader. They are also able to have various experiences to realize the field of ODA and NGOs' activities in a developing country that has different cultures and customs. They are thus able to deepen their understandings of various structures and people in international society and to think about how a leader should be.

Through the OIP, students are able to have the opportunity to consider effective policy and ideas to utilize their specialty for the society by acquiring the wisdom to judge and act as a leader.

15 The Role of International Organizations: Solving Complex Global Issues

Eriko Ishida Kawai

Numerous complex global issues have emerged worldwide, such as global warming and refugee problems. International organizations have been playing important roles in solving many of these problems. In this chapter, the current and future role of international organizations will be analyzed. The fact that cooperation and coordination among international organizations, research institutes, companies and non-governmental organizations (NGOs) are becoming increasingly important will be discussed. The expected changes to the roles played by international organizations are also mentioned, as is the potential of supranational organizations.

Role of international organizations in global society

The number and types of worldwide, large-scale problems, such as climate change and financial crisis, are on the rise. They widely affect many nations and cannot be solved through the efforts of one country alone. Therefore, the importance of international organizations is increasing more than ever before. One of the roles of international organizations is to provide a place or forum where governments and NGO's can gather to discuss global issues. Various significant world conferences have been organized by international organizations in recent years, such as the Earth Summit, the Millennium Summit, the World Summit and the COP (Conference of Parties to the United Nations Framework Convention on Climate Change).

The UN held the Earth Summit in Rio de Janeiro in 1992. The focus of the summit was sustainable development with the aim of conserving the environment. A total of 175 governments participated, with 116 of these sending their heads of state. Some 2,400 representatives of

NGOs attended and 17,000 people participated in an NGO 'Global Forum' held in parallel with the Earth Summit.

The United Nations Millennium Summit was held in New York City in 2000. The focus of the summit was on various global issues such as poverty, AIDS and the equitable distribution of benefits arising from globalization. One hundred and eighty-nine member states of the UN agreed to help citizens in the world's poorest countries to achieve a better life by 2015. The Millennium Development Goals (MDGs) were set and the framework and progress outline was established.

Ten years after the Earth Summit in Rio de Janeiro, the UN held the World Summit on Sustainable Development in Johannesburg in 2002, and many business leaders and NGOs participated.

Among the important roles of international organizations are the gathering of accurate statistical data and its evaluation, as well as the assessment of scientific data and its dissemination. For example, in 2007, the Intergovernmental Panel on Climate Change (IPCC), a UN organization specializing in the evaluation of research on climate change, played an important role in discussions on climate change, and was awarded the Nobel Peace Prize jointly with former US Vice President, Al Gore.

The IPCC was created by the United Nations Environmental Program (UNEP) and the World Meteorological Organization (WMO) in 1988. It consists of a network of scientists volunteering their services to scientifically evaluate and assess data while maintaining a politically neutral stance. Initially, opinions about the causes of climate change were divided, even among scientists. Many believed that the causes and possible outcomes had not been scientifically proven. This being the case, the IPCC collected, examined and analyzed data using the latest climate models, and evaluated scientific papers. They came to the conclusion that, with high probability, global warming was in progress. Climate change has become a scientifically proven fact, and the IPCC is raising awareness among member states.

Another important role of international organizations is coordinating nations, establishing international regulations and creating a world order. International organizations have established many international rules that impact on the survival of humanity in areas such as the environment, food production and sustainable development. They fill a void with respect to regulating the international public good, where government regulations have fallen short, and have integrated these regulations with international rules.

The issues often discussed at conferences concern how international organizations might impose country specific regulations. Let's take the financial industry as an example, where globalization is particularly advanced. In order to avoid a chain of global financial crises, the Financial Stability Board and the Basel Committee on Banking Supervision at the Bank for International Settlements are at the center of creating and imposing international financial regulations. The so-called Basel Accords, regulations concerning banks' capital ratio that aim to prevent the bankruptcy of major banks involved in international business, are one such example. They are a 'gentlemen's agreement' established by the banking supervisory authorities of the major countries. Rather than imposing legal penalties on banks that do not comply with these regulations, it works through peer pressure. When certain banks do not comply with the regulations, they may face reputational risk. Trust is lost and this makes banking operations difficult.

Since international rule-making is directly linked to a specific country's interests, agreement is often extremely difficult. For example, constraints on greenhouse gas emissions may depress economic activity. Even when regulations are adopted, a system for ensuring compliance does not exist. It is a major challenge for international organizations to reach agreement, and then to ensure compliance.

The work of international organizations involves coordinating nations with different views and interests. As mentioned above, while their work transcends the domain of a single nation, they do not have the authority to supersede national sovereignty. Usually, resolutions and treaties are instituted by international organizations such as the UN. Adoption of these treaties is, however, left to national discretion.

The European Union (EU) is an exception and is the world's first supranational international union transcending national boundaries. The EU was established through a mutual agreement among sovereign nations; however, it has the power to constrain member states through regulations.

EU laws and regulations apply to the citizens of all member states. As a result, policies in the region are standardized under European Commission law. Regulation constraints fall under two categories:

directives and *decisions*. The distinction lies within the scope of their legal binding force. Once a general agreement is made, even if a country opposes a regulation, it must still comply. With an issue such as climate change, in coming together as a single negotiating entity, the EU has succeeded in showing a strong presence.

What are international organizations?

International organizations can be defined as multi-national organizations that are established, based on a treaty, to help to achieve common objectives among nations. International organizations are also sometimes referred to as inter-governmental organizations. In order to carry out their missions and mandates, they include internal entities. Taking the UN as an example, there is the General Assembly at which all member states participate, the Council composed of representatives of some member states and the Secretariat, involved in the preparation of conferences and the execution of decisions made by the Council and the General Assembly.

Up to 238 international organizations of various sizes exist, according to the International Organization Yearbook (2003–4). The UN is one of the most important ones. Established as a result of two world wars, its mission is centered on peacebuilding and security. Other international organizations play major roles with respect to development issues, the fight against hunger and poverty and more recently, environmental problems.

The principal organs within the UN are the General Assembly, the Security Council, the Economic and Social Council, the International Court of Justice and the Secretariat. There are also related programs and funds of the General Assembly, managed by entities such as the United Nations Development Program (UNDP), United Nations Environmental Program (UNEP) and the United Nations Children's Fund (UNICEF).

There are fifteen specialized agencies within the UN, including the United Nations Food and Agriculture Organization (FAO), the World Health Organization (WHO) and the United Nations Educational, Scientific and Cultural Organization (UNESCO). The International Civil Activation Organization (ICAO) and the International Telecommunications Union (ITU) are responsible for managing technical issues involving standards and criteria. Each agency is linked to the UN through a special agreement, but they are independent organizations that report directly to the General Assembly and the Economic and Social Council.

The World Bank Group and the International Monetary Fund (IMF), which were established by the Bretton Woods treaty, are also specialized agencies. They deal with international finance particularly with respect to financing in developing countries, and although they are part of the UN system, their link to the UN is weak (see the United Nations Organization Chart on page 264–265).

There are also prominent international organizations that are not part of the UN family. These are organizations such as the World Trade Organization (WTO) and the Organization for Economic Cooperation and Development (OECD), the latter of which has thirty-four members from developed countries. The mission of the OECD has been to contribute towards economic growth, trade liberalization and development assistance. Recently, new areas such as the environment, sustainable development and governance have been added. There are about thirty committees active in collecting data and analyzing the policy issues of member countries.

The WTO was created in 1995, after the Uruguay Round Negotiations, which resulted in many significant developments with respect to trade rules, such as lower tariffs. The WTO is now aiming to implement the new Doha Round Negotiations, however there were difficulties in reaching any agreement. As achieving any worldwide agreement can be very difficult with so many different interest groups involved, regional trade agreements such as the Trans-Pacific Partnership (TPP) are expected to play a larger role.

Furthermore, there are many new international organizations active in limited regions, such as the Asian Development Bank.

Challenges faced by international organizations

International organizations face a wide array of problems that often involve a lack of expertise and knowledge, human resources and funds. As a result of the worsening of the financial situation of member countries, lack of funds has become a major problem. The UN's regular budget for administrative expenses and staff payroll amounts to $5.5 billion (biennium budget 2016–17). In addition to the regular budget, there are voluntary and assessed contributions, used for development aid and peacekeeping operations. These contributions total about $40 billion for the entire UN.

In order to secure funds, significant time and effort is spent in negotiating financial contributions from member states.

Because international organizations depend on the contributions and dues of member states, there is a need to carry out activities in line with their intentions. For example, in 1984, the US left UNESCO. At the time the US accused UNESCO's activities of being 'politicized', and thus decided to discontinue their membership. The UK also followed suit. In 1999, the Japanese diplomat Koichiro Matuura became UNESCO's Secretary-General and committed to repairing the relationship between his organization and the US, as well as introducing better governance. Thanks to his efforts, in October 2003, the US resumed its membership. However, since Palestine became a formal member of UNESCO, the US has frozen its contributions. UNESCO is currently facing a budget crisis and cost cutting as a result of US non-payment. The US was formerly responsible for 22% of its budget.

The second problem faced by international organizations in the past was having adequate human resources. The most problematic and common occurrence with respect to human resources was favoritism. In the past, some people were being employed without going through the standard recruitment process. Some received political appointments because of their nationality or gender. Appointments for reasons other than qualifications and individual capability have become a source of great frustration and inefficiency. In contrast, at organizations such as the World Bank and the OECD, these problems occur less often. Individuals are selected based on their qualifications and skills, regardless of gender or nationality.

The third problem has to do with organizational issues. Sometimes 'turf wars' occur because of an overlap of the mandates of two or more institutions. Cooperation and collaboration are often dysfunctional. For example, the WHO, UNICEF and the United Nations Population Fund (UNPF) are all part of the UN family and are all involved in medical care and public health. However, cooperation between these institutions does not go smoothly. The same thing can be said about coordination among special agencies. For example, in order to solve problems related to poverty, it is necessary to work across multiple areas. This means working on trade, the environment, education, and at times, simultaneously, on civil war prevention. Good coordination is necessary to improve and enhance cooperation.

Incorporating business skills such as management and IT may increase efficiency. Member states also often point out that because these organizations have become so large, they have become more rigid and often inflexible, resulting in bureaucratization. This also affects ethics and motivation. To combat inefficiency and the lack of available skills, the use of HR consultants and others from outside the organization has increased.

Finally, it is important to be critical and to pay attention to accountability and governance. For example, there has been significant criticism concerning the bureaucracy of the enormous EU Secretariat. EU bureaucrats make numerous rules without considering national differences and exert enormous power, sometimes greater than that of an entire small nation.

International organization-civil society partnerships

At the Millennium Summit in 2000, former UN Secretary General Kofi Annan pleaded for the need for partnerships with a variety of actors such as NGOs and private corporations: 'The international public domain – including the United Nations – must be opened further to the participation of the many actors whose contributions are essential to managing the path of globalization' (from a plenary titled 'We the Peoples: The Role of the United Nations in the 21st Century').

In recent years, the relationship between many NGOs and the UN has become very close, resulting in numerous collaborations. When working on common challenges, it has become clear that without the participation of NGOs representing civil society, it is difficult to attain meaningful results. Some NGOs have been able to register at conferences held by UNESCO. They are invited to attend the advisory council and allowed to speak at certain conferences. NGOs involved in public relations partner with the UN Department of Public Information (DPI). The UN also invites NGO representatives to its regular briefings.

Under such partnerships, NGOs are now participating in a number of international conferences. For example, at COP11 in 2005, the total number of participants was approximately 9,500 people. Of that total, some 2,800 were official government representatives, and 5,400 were NGO members with observer status. Although NGOs have no voting or negotiation rights, they do have a voice. They aim to influence intergovernmental negotiations through informal channels via side meetings.

NGOs are gaining more and more importance through their growing financial power and expertise. In 1995, government aid carried out by NGOs represented a mere 4.6%; by 2004 it had increased to 13%. NGO activity in Japan is also on the rise. Currently, according to the Minister of Foreign Affairs, there are more than 400 Japanese NGOs involved in international collaborative activities.

Another positive trend is the increase in the number of foundations created by private donors. These help finance the UN, which as noted above has been suffering from lack of funding. For example, Ted Turner, the founder of CNN (a global news broadcasting service) has donated more than one billion dollars to the United Nations Foundation[1], established in order to assist in funding its activities.

There is also a growing role for international organizations in coordinating the activities of various actors. For example, with regards to the Polio Eradication Initiative, a partnership was created between the WHO, UNICEF, Rotary International, the US Center for Disease Control and the Bill & Melinda Gates Foundation, which resulted in a several hundred million dollar donation to the polio eradication program. Furthermore, public health ministries, governments, foundations, corporations and volunteers participated in this campaign. In 1988, polio was present in 125 countries, and 350,000 children were affected every year. Due to the polio eradication campaign, 2.5 billion people were successfully vaccinated. The number of countries where polio continues to exist has been reduced to two: Pakistan and Afghanistan.

Towards more cooperation with business

As the relationship between the UN and civil society becomes closer, so does its cooperation with the business world. There has been a shift from the direct regulation of businesses toward establishing general business regulations by establishing guidelines and rules for the smooth functioning of business. Previously, as outlined in the OECD's Guidelines for Multinational Enterprises, emphasis was placed on regulating the business activities of multinational corporations. Many multinational companies have immense financial power and influence, greater than that of some of the developing countries. Presently, the focus has shifted to regulation in order to promote business activity.

The United Nations Global Compact[2] was proposed by former UN Secretary-General Kofi Annan and was launched in 2000. In a nutshell, it concerns corporate social responsibility (CSR). It encourages global businesses to adopt socially responsible and sustainable policies. Companies are part of civil society, and therefore, they should fulfill their responsibilities as global citizens through good corporate governance. There has also been rapid growth in Socially Responsible Investment (SRI), as investors, such as pension funds, realized that companies with better CSR and governance tend to outperform the market in terms of long-term growth, and also have a better image.

Within a company's range of influence, both companies and investors will support a series of essential values, ranging from human rights, labor and environmental issues, to combating corruption. These values fall in line with the Universal Declaration of Human Rights, the Declaration on Fundamental Rights of Labor, the Rio de Janeiro Declaration on Environment and Development and the Millennium Development Declaration.

Currently, more than 10,000 companies of 145 countries have signed and are participating in the United Nations Global Compact. In January 2015, the number of Japanese companies participating in the Global Compact Japan Network was 192, and it is growing.

Collaboration with academia

Through a program called Academic Impact[3], the UN also collaborates with higher education institutions around the world. This program also encourages these institutions to cooperate among themselves. By incorporating the UN's Millennium Development Goals, they have created a mechanism through which they have been integrated into UN operations and activities.

The OECD, for example, hires researchers as consultants to carry out joint research and uses their network for organizing conferences.

The WHO has an extensive network of medical professionals they can reach out to with specialist knowledge related to specific diseases. WHO officers with PhDs in public health do not have this specialist knowledge, and hence it is important for them to work with medical professionals outside of their organization.

Kyoto University has taken part in a variety of activities that have involved cooperation with international organizations. For example,

UNESCO and Kyoto University signed an internship agreement for graduate students in 2012. Students take part in programs such as the UNESCO International Hydrological Program. This has led to the establishment of UNITWIN (University Twinning and Networking) at the Kyoto University Disaster Prevention Research Institute. There, students take part in the Rio de la Plata river basin workshop.

In addition, the United Nations Environment Program (UNEP) and Kyoto University signed a cooperation agreement in 2013, supporting graduate students working on clean water management. Over the past ten years, both organizations have conducted environmental assessments on a global, regional and national scale. Through mutual cooperation and the deployment of graduate students and experts, they created an international report called Global Environmental Monitoring Systems (GEMS/Water). Currently, they are involved in the United Nations World Water Assessment Program (UNWWAP) and Sustainable Development Goals (SDGs) as experts in the field. Furthermore, the United Nations Food and Agriculture Organization (FAO) signed an internship agreement with the Graduate School of Advanced Integrated Studies in Human Survivability at Kyoto University in the areas of microbiology and crop yield models.

Innovative financing

As a solution to the problem of the limited financial resources of international organizations, various innovative financing methods and mechanisms have been tried. The Leading Group on Innovative Financing for Development[4] was founded in Paris in 2006. It constitutes an attempt at securing new sources of revenue and does not rely on member country contributions, as is the case with the international solidarity tax.

UNITAID[5] was established in 2006. It was founded as an attempt to make various medical treatments, such as those for malaria and HIV, more readily accessible in developing countries. A solidarity tax on air travel is its main source of funding. In creating a solidarity tax, a tax on the purchase of airplane tickets, income is redistributed from the privileged to the poor. It was initially adopted by the five founding countries, France, Brazil, the UK, Norway and Chile, at a conference in Paris in 2005.

As a new way of governance of international organizations, the Board of UNITAID does not only represent its founding countries, but also includes representatives from the WHO and civil society. As UNITAID has also received funding from other foundations, there have been demands for full accountability regarding its activities. To that end, the organization is striving to secure third-party evaluation in order to ensure transparency.

Reflecting on the global financial crisis that began in 2008, active discussions have been ongoing regarding a new system called the Financial Transactions Tax (FTT), a mechanism through which certain financial transactions could be taxed. As a result of the FTT, total transaction costs would become higher; therefore, short-term, speculative transactions may be suppressed. This taxation system was adopted at a meeting of the finance ministers of eleven European countries on January 22, 2013. However, the UK, a financial powerhouse as far as EU member states are concerned, opposed its introduction and filed suit against it in the European Court of Justice. As a result, the introduction of this new system has been delayed. Tax rates are expected to be lower if it is implemented.

To this point, such a tax has not been considered or discussed in countries other than the EU. Since financial markets are global, should this new tax only be adopted in the EU, market players such as hedge funds will trade outside the EU and the tax will be less effective. However, if this type of tax was implemented by the world's major countries, it would result in huge tax revenues. According to a 2007 estimate with a tax rate of 0.01%, tax revenues could amount to roughly $286 billion.

In comparison to the UN regular budget of $5.5 billion, as described earlier, the impact of this tax would be huge. If part of the tax were used for resolving global environmental and economic development issues, it would help realize many projects that currently cannot be implemented due to a lack of funding and political will. That would be great news for many international organizations. If international organizations did not have to rely on member countries' contributions, they could promote scientifically and economically proven policies without having to deal with political pressure and the political and business interests of individual countries.

Future challenges

In this chapter, the role of international organizations in peacebuilding was not discussed. There are many problems, such as the situation in the Ukraine and Syria, as well as the growing power of the so-called Islamic State, that pose a threat to the international community. The United Nations Security Council Committee's actions are dysfunctional as far as these problems are concerned. Furthermore, the US and Europe seem to be, at this point in time, politically inward-looking and focused on their own national interests.

The existing international financial order has been challenged by emerging economies. In July 2014, the New Development Bank was established by five emerging countries: Brazil, Russia, India, China and South Africa. Its headquarters are located in Shanghai, and its first president is of Indian origin. It is sometimes referred to as the BRICS Development Bank for Infrastructure and Sustainable Development. Furthermore, China founded the Asian Infrastructure Investment Bank (AIIB) in October 2014, and is its leader. These new development banks will be new forces within the international financial system in the future. They are neither western-driven nor centered on the IMF or World Bank. The world is monitoring their development closely.

The old concept of 'developing countries vs. advanced countries' as previously formulated by the UN, no longer applies to reality. Responsibility with regard to the emissions of developing countries was on the agenda of COP21 held in Paris, 2015. It is no longer possible to put the entire responsibility for emissions on developed countries. China and India have undergone rapid economic growth. They are, respectively, first and third amongst countries with the highest emissions. It is also incorrect to categorize them as 'developing countries'. Their level of economic and political development, as well as their economic influence, is very different from that of 'developing countries'. As the BRIC emerging countries are gaining significant power, they are not bearing their share of responsibility as part of the international community.

Summary

The roles of international organizations are changing over time. Problems that transcend national interests, so-called 'global public goods' such as the environment, require international organizations to play the role of coordinating between nations. Establishing international regulations is also necessary, as multinational companies operate across borders. Although an agreement has been reached with regard to emissions regulation, forcing countries to comply remains difficult. Even after the establishment of the Kyoto Protocol in 1997, an effective framework for fighting climate change does not exist.

Global issues, such as environmental and sustainable development problems, need to be addressed appropriately. What is necessary is not only a top-down governmental approach, but also a bottom-up tactic. In the past, arbitrating nations' differing interests was one of the major roles played by international organizations. Decisions were made through top-level national negotiations. In the future, international organizations are expected to take on a coordinating role – helping to increase cooperation between governments, NGOs, university research institutes and businesses. Taking development aid as an example, many NGOs are presently acting on their own, creating great inefficiency due to the overlapping of activities and lack of coordination with host countries. In response to this inefficiency, the UN has been increasing field officers in regional offices instead of those located at headquarters.

International organizations are subject to the financial constraints of member states, which limit their activities. In the future, it is conceivable that a new global tax, similar to the solidarity tax on airline tickets, may become an active source of funding. These funds could be used to tackle global challenges such as climate change and poverty.

The accountability of international organizations and the quality of their governance have been questioned. To receive continuous support from member states and civil society, they need to avoid inefficiencies and ensure that transparency is guaranteed in their operations.

With respect to international organizations' wide range of responsibilities, we must bear in mind the development of new financial institutions and emergence of new economies. China is a prime example. It is therefore important to watch carefully, as a new international economic regime may be about to be born.

16 Risk Management

Takashi Kanamura and Koichiro Oshima

Introduction

Risk management of social and economic activities is one of the most important issues that highly specialized contemporary society faces. Entities, including companies, are always vulnerable to social and economic risks stemming from future uncertainties. Risk is written in Chinese characters as 危機, which includes two contexts of danger 危険 and opportunity 危機. After understanding risk profiles from various perspectives, entities have thus far had to manage social and economic risks, categorized as those to be hedged or taken. These result in costs from potential dangers or returns from potential opportunities. However, there has been a gradual realization that it is difficult to find ways to solve social and economic risks in modern society by simply specializing or discretizing the categorized risks. For example, utility companies may consider the risk posed by nuclear power plants in terms of cash flows, while the potential for emergency situations may not necessarily be taken into account, in that the risk is not accepted by reinsurance. That is, existing risk categorization cannot solve issues related to risk. Additionally, it is myopically considered that the risk of climate change may be hedged using an emissions trading scheme, but in reality it is thought that the risk posed by climate change should voluntarily be seen as a technology innovation opportunity, resulting in actual emissions reduction in the mid- or long-term. It is safe to say that finding the solution to risk-related issues is significantly restricted by breaking the risks into parts. As the opposite vector to breaking down the risks, it has been gradually recognized that we need to solve the related issues by understanding segmented areas of expertise and connecting them flexibly in response.

This section attempts to discuss the problem of risk management through the new field of Human Survivability Studies (HSS) by going beyond the various areas of expertise. In this sense, the approach completely differs from that employed by existing disciplines. This may be seen as substantially disconcerting from the perspective of existing expertise. However, the new framework reorganizes different areas of expertise in order to solve risk-related problems and identify new values from the solutions. That is, we hope that our project can be considered as an experimental and completely different approach from existing expertise.

You may question the difference between existing Sustainability Studies and HSS. According to a dictionary, 'sustainability' derives from the notion of support from below in Latin, resulting in the context that existing prosperity maintains the world in the future. It follows that Sustainability Studies may involve the extension of existing expertise. In contrast, HSS is based on the word 'survivability', which stems from the Latin meaning 'a life beyond everything'. By transcending the world's current prosperity, it includes finding solutions to global complex issues and continuing to live in light of these solutions. The new approach of HSS entails a serious attitude toward life including the message: 'we must live here, continue to live and manage to live'. Thus, we can safely say that HSS can offer much needed expertise in order to achieve the ultimate goal of the continuance of human life through interdisciplinary engagement. How to transcend existing disparate areas of expertise and how to reorganize them are the keys to this approach and will enable us to solve the big issue of how to continue to exist. In this sense, HSS is transcending disciplinary boundaries by following the idea of 'in the same boat'.

The development of HSS is not aligned with existing sustainable development as outlined by the World Commission on Environment and Development (WCED) in the 1980s, but is instead connected to the new concept of 'survival development' in the sense of humanity's struggle to continue to live. In addition, the framework of existing fragmented areas of expertise implies the mechanization of human beings, in which human beings strive to achieve goals including solutions to global complex issues using microscopic views. However, it is often said that machines will not be able to take the place of human beings in the future. Human beings work to achieve goals including problem solving using intuitive and comprehensive recognition, which is quite difficult for robots. Like this, HSS, which reorganizes existing distinct areas of expertise to solve problems, can imply the mean reversion to the nature of human beings. In concrete terms, this new approach takes risk management issues on Earth as its starting point, and clarifies the essence of these problems. Then, by breaking existing frameworks of expertise into parts and reorganizing them according to risk management issues in a transdisciplinary way, studies using this approach attempt to demonstrate the possibility of finding solutions to them. HSS, which deals with solving practical issues and value creation, and existing areas of expertise differ in quality, but the former is not superior to the latter.

The next section examines three examples of risk management issues regarding business, natural resource development and science and technology ethics. The necessity of HSS is confirmed in the context of these three concrete practical issues after the limits to solving them using existing approaches are outlined.

Risk management in business activities

Risk management in business activities is a serious issue that concerns the viability of companies. A recent development in this field highlights the importance and urgency of risk management and calls for the assignment of a chief risk officer (CRO) to the management board to be responsible for dealing with risk management issues in practice. CROs work hard to keep companies going concerns by monitoring risks in complex and various business activities and conducting companywide risk management assessments. This subsection discusses three risk management issues: business decision-making that concerns the viability of companies, financial trading related to threats to the existence of financial institutions and project execution regarding company value in terms of cash generation from company projects.

Regarding risk management issues in business decision-making, we take the example of a stably growing power and gas utility company that enters into energy trading and engages in risk management by betting on energy trading with the aim of company survivability. A utility company that conducts its businesses with stable growth in the long run may try to avoid the market risk of energy trading, which is difficult to predict. This is due to the fact that a utility company can achieve high credit ratings based on its corporate management and can generate returns from the credit risk of energy trading. In concrete, a utility company may try to enter into risk neutral energy trading that matches short and long positions, resulting in a net profit from reliable price spreads based on the credit worthiness that avoids the market risk of energy trading. In order for a utility company to position energy trading as a new pillar of its businesses, the chief business executive who oversees the running of the company is not only familiar with the marketing of energy trading but also must be able to supervise the entire business comprehensively from above, including having a full understanding of risk assessment and management and the energy trading legal framework such as the Dodd-Frank law, and also have deep knowledge of energy trading.

Regarding marketing, the chief executive needs not only technical skills to enable deep insight into energy market observations and analyses, but also social and human science skills to allow them to negotiate with counter parties used for top-down sales. In addition, in order to determine the risk limits of energy trading and the direction of the energy trading business, the chief executive, in a broad sense, needs natural science skills to understand quantitative risk assessment and management. Furthermore, they are required to have legal skills to judge the validity of energy trading in the context of new legal frameworks that have been implemented in the wake of recent financial turmoil.

Another necessary trait is a philosophical framework to govern the direction of the company and relate to broader society when energy trading becomes a new pillar of the company. In this regard, when a top business executive works to solve business decision-making risk management issues and expands their new business for the survivability of their company, the more segmented and specialized the business components become, the more strongly the executive needs to possess the ability innate to HSS to understand and connect the fragmented areas of expertise in order to execute the business practically, i.e., make decisions. That is to say, existing specialized and segmented approaches in academia cannot solve the risk management issues in decision-making related to a company's survival and cannot create new value.

The risk of trading in financial sectors is represented as the extent to which future cash flows deviate from expectations, using the word 'finance'. In short, the deterministic future loss is not considered as a risk. Assuming that future uncertainties can be expressed quantitatively, the degree of variation from the expectation, e.g., the standard deviation for normal distribution, is taken as the risk. If the distribution is not symmetric like a normal distribution but skewed like a lognormal distribution, the risk should be adjusted depending on the shape of the distribution.

Categorization is the typical starting point in the process of managing risks arising from financial trading represented by the deviation from expectations. It is well known that two types of risk exist in financial trading. One is referred to as 'market risk', which represents changes to profit and loss caused by market volatility. The other is referred to as 'credit risk', representing existing position changes caused by counterparty default. Market risk is mainly measured by Value at Risk (VaR), which assumes that future asset price returns follow a normal distribution and represents profit and loss fluctuation of trading positions using a standard deviation of a normal distribution. Here, a confidence level such as 99% is taken depending on the tolerance of financial institutions to risk. Credit risk is measured by a restructuring cost due to counterparty default. In concrete, it is calculated according to the maximum exposure that can occur in the future. In addition to the two risks outlined above, operational risk (oprisk) is now highlighted as the third risk in financial trading. In financial trading, this is the other type of risk in which inappropriate or malfunctional internal processes, or human, systemic or exogenous events cause losses. Oprisk is quantified using the total sum of future losses driven by these factors. Thus, oprisks can be considered as those other than market and credit risks. However, there is a significant issue for the application of oprisk to financial trading businesses that differs from existing market and credit risks. That is, existing segmented areas of expertise such as the deeply specialized finance sector cannot provide answers to the quantification of oprisk, even when existing expertise attempts to qualify it. For example, input mistakes in financial trading, one of the significant oprisks, have to be mathematically modeled taking into account the structure of human behavior depending on human psychology. That is, in addition to statistics, different areas of expertise including psychology should be interconnected. More seriously, the management quality of oprisk by financial institutions has the potential to become a threat to the survival of financial institutions. While the concept of oprisk related to the continuation of financial institutions is understandable, the detailed quantification of such risk is still in progress. In order to solve risk management issues in financial trading, HSS, i.e., transdisciplinary fields, are required.

Finally, let's examine risk management in project execution directly related to the viability of a company. As the basic knowledge underpinning project management, the US Project Management Association proposed the Project Management Body of Knowledge. In this framework, a project management process includes risk management planning, risk identification, risk qualification and quantification, risk response planning and risk control.

Risk management planning defines the detailed implementation of project risk management. Risk identification highlights the determination of high impact risks and the detailed description of risk characterizations. In order to conduct further analyses, qualification analysis prioritizes risks by taking into account the occurrence of probabilities and their impacts. By doing this, we can focus on more prioritized risks and reduce future uncertainties. Quantitative risk analysis then calculates the impact of each risk on the project as a whole. By representing risk with a distribution, sensitivity analysis is conducted using, for example, tornado charts. In addition, risk response planning is carried out to identify the options and actions that can clarify the purpose of a project and alleviate potential threats. By doing this, we can handle each risk depending on priorities, budget planning, schedule and project management planning. Finally, risk control enables risk response planning to be reliably executed by observing time dependent risks and can comprehensively evaluate the efficiency of a project's risk process by tracking identified risks and monitoring residual ones. This risk control process can improve the efficiency of the risk approach, which continuously optimizes risk response planning through the project's life cycle.

Project management may specify the risk management process in detail and successfully conduct risk control. However, we have to remember that projects are conducted and lead by fallible human beings, not automatic machines. Therefore, even if risk management manuals are always revised properly, project managers have to navigate in the right direction in order to ensure the project will survive by taking into account the balance of the entire project, by identifying the essence of the problem from the perspective of risk and by recognizing and solving the problems by reorganizing different areas of expertise if unexpected events occur. That is to say, risk management in project execution cannot solve the issues by breaking things into component parts, i.e., using existing expertise in order to control risks practically. Risk management in project execution also needs the new concept or 'framework' of HSS, which reorganizes different areas of expertise and connects them in order to solve risk management issues related to human survivability, creating new value from risk management in project execution.

Risk and its management in resource development

Currently, we rely heavily on four kinds of primary energy sources – coal, oil, natural gas and nuclear power. These are all limited in amount, non-renewable and face depletion risk. On the other hand, the existence of large scale undiscovered oil fields is either confirmed or expected. For example, it is said that there are vast quantities of undiscovered resources under the Arctic Ocean. Although searching and mining the ocean floor is not easy, such difficulty can be overcome by innovation. In the US, moreover, it is reported that there is enough shale gas contained in the soil to supply that country for 300 years. In Japan's case, it has been reported that methane hydrate exists under its territorial waters. Research and experiments have been conducted in order to extract methane hydrate safely. Some people remain optimistic about energy problems because of these potentially available resources. Can we, however, realistically be optimistic? The main component of shale gas and methane hydrate is methane gas, which is the main component of natural gas, and thus they cannot be regarded as new primary energy sources. In other words, they belong to the abovementioned conventional primary energy sources. Humankind is set to use up four kinds of natural resources mentioned above within a few hundred years, which Earth has saved for 4.6 billion years since its birth. If we can discover a new unlimited energy source, there would be no problem with depleting these resources, but in reality, this is clearly not the case (see Chapter Thirteen for perspectives on new forms of energy). While leaving the discovery of a fifth energy source to the experts, I'd like to discuss the development of shale gas and methane hydrate, and examine the accompanying risks. Additionally, I attempt to explain the development of renewable energy (such as hydraulic, wind and geothermal power and so on) and the concomitant risks. Finally, I touch upon the importance of the approach enabled by HSS.

Shale gas is contained in fine-grained rock called shale. It was impossible to extract shale gas using conventional methods, until George Mitchell developed new techniques called horizontal drilling and hydraulic fracturing, which can be used to make cracks in shale rock and efficiently extract the gas within. Industrial production and usage of shale gas has already begun in the US. Methane gas itself, however, remains problematic, because its greenhouse effect is twenty times as strong as that of CO_2. Shale gas is quite an efficient fuel, but at the same time, it is concerning that leaked residual methane gas, if left unburned, may cause detrimental effects on climate change. Moreover, there are concerns regarding the possibility of environmental pollution caused by chemicals used in the process of hydraulic fracturing, and also the potential induction of earthquakes in relation to the injection of copious amounts of water underground. Risk and its management around the development of shale gas must be addressed through not only the field of engineering, but also via a multifaceted approach including measures to prevent climate change and environmental disruption as well as legislative measures.

Methane hydrate is a dodecahedron crystal with pentagonal faces, each of twenty apexes of which has an oxygen atom originating from a water molecule. At the center of the structure sits a methane molecule. Methane hydrate exists in a solid-state in the water below 400 meters below sea level. Once extracted from the sea, however, it becomes sherbet-like. If ignited, it burns with a blue flame. It is estimated that there is six trillion cubic meters of methane hydrate deposits in the oceans around Japan, and 250 trillion cubic meters of it in the world. Although methane hydrate seems a potential alternative energy source for the future, there are concerns similar to those regarding shale gas development: the potential induction of earthquakes and detrimental effects on the environment. In addition, the actual utility of methane hydrate depends on whether we can come up with an innovative and economically plausible technique. As in the case of shale gas, we need the perspective of HSS in order to resolve the risk and management problems around methane hydrate.

Next, I'd like to touch upon the development of natural energy and surrounding risks. Even though natural energy resources such as wind, geothermal and solar are deemed 'clean', they do not come problem free. In terms of wind power generation, Japan has a prima facie advantage, as it is surrounded by sea; Japan's territorial waters, including its exclusive economic zone, covers 4.48 million m², which is the sixth largest in the world. Although some areas are not suitable for offshore wind power generation, it is estimated that such method can generate as much as 600,000 MW (equivalent to 230 nuclear power plants), combining fixed bed type and floating type turbines, even if we narrow down the areas of operation according to certain conditions, such as 7.5 m in annual average wind velocity, or a water depth of 200 m or shallower. While research and experiments are ongoing, there are problems in terms of noise pollution from wind turbines and fishing rights. In contrast, European countries with shallow surrounding waters, such as the UK, have already begun operations of fixed bed type offshore wind power generation. When it comes to geothermal power generation, Japan has land use restrictions and faces legal obstacles. For example, power plants cannot be built in national parks or semi-national parks.

Solar panels haven't changed much in terms of size over the past twenty years, as the light energy conversion efficiency has not improved. Therefore, we still need a vast amount of land to house solar panels. Indeed, in order to produce the same amount of electric power generated by a natural gas power plant, we need 3,000 times as much land to accommodate the solar panels. If we try to generate enough electricity for the whole country solely using solar panels, we would need 25,000 m² of land, which is equivalent to Shikoku Island and Ibaraki Prefecture combined. Construction cost is also an issue, as building a panel that can generate one kW of electricity costs 500,000 yen.

Finally, I'd like to touch upon problems with biofuels. One example of biofuel is ethanol produced from corn. Corn contains a large amount of starch. Starch can be decomposed into glucose using enzymes, and glucose can be made into ethanol when fermented. Using this method, we can obtain ethanol with a concentration of up to 20%. To increase the concentration, ethanol has to be separated from water. This necessitates a significant amount of energy. Since large amounts of fuels are indispensable in the production, transportation and fermentation of corn in the first place, it is a concern that energy consumed in the process of making biofuels might exceed the amount of energy gained. Moreover, food prices soar when food resources

– corn – are converted into fuel. New technology that can convert inedible cellulose into ethanol is desirable.

As explained above, in order to manage the risks regarding natural energy sources, such as wind, solar and biofuels, it is necessary to have a comprehensive approach involving multi-dimensional aspects, including not only engineering, but also law, the economy and so on. Once again the need for HSS, which maintains a transdisciplinary approach, becomes evident.

Risk management and science and engineering ethics

Aside from associations of government-recognized occupations, such as medical doctors and lawyers, there has been no code of ethics in Japan, because it was presupposed that Japanese people would voluntarily stick to the general framework of ethical rules. Concepts like science or engineering ethics were not considered until recently. In other words, we may have been fortunate in not having to be conscious of ethics. In the US, textbooks about engineering ethics have been available for decades, whereas there were none until recently in Japan. It is not surprising, therefore, that engineering students at universities did not receive education in ethics. This is because Japanese people's strong sense of belonging to corporations places more obligation on employers than the public.

However, since the Tokaimura nuclear accident – caused by negligence regarding the predetermined procedure to melt uranium compound – occurred in 1999, there has been a series of incidents in which the ethics of engineers or companies have been questioned. These include:

1. The Snow Brand Milk Products Co., Ltd. food poisoning accident, caused by a lack of basic sanitation management of food products, resulting in the poisoning of more than 10,000 people.
2. The attempt to conceal issues at several nuclear power plants, where no record was made that these problems existed and that repairs were needed, and on the day these issues were discovered the facts were repeatedly manipulated
3. The West Japan Railway Company Fukuchiyama-Line derailment accident, which occurred on a curve due to excessive speed.

Because of these incidents, individual engineers are now required to be fully aware of the ethical aspects to their work. How engineers should deal with work-related ethical problems is a matter for discussion. It is certain that all the engineers involved had to face considerable difficulties in the wake of the 2011 accident at the Fukushima-Daiichi nuclear power plant, caused by the Great East Japan Earthquake. At the same time, a single engineer cannot be responsible for answering questions like 'whether or not nuclear power generation should be suspended'. Rather, such issues must be discussed by various stakeholders including engineers, and encompass diverse opinions from multiple areas, before reaching a conclusion. Many universities have now initiated ethics education. I hope it will become a compulsory subject in the near future.

Recently, frequent wrongdoings in the course of scientific research have come to light. These incidents disturb the mutual trust that is the major premise of any scientific research. Case studies can be categorized into three types: (1) fabrication, or creating fake data; (2) falsification, or intentionally modifying/misrepresenting relevant data; and (3) plagiarism, or stealing other scientists' ideas. The fabrication and falsification of papers on embryonic stem cells, cold nuclear fusion and organic high temperature superconductors are still fresh in our memories. In addition, many cases of data modification or fabrication through the abuse of image processing technology have been reported. Such situation was made manifest against a background of tremendous progress in the image processing capability of computers. We live in an era when a person's photo can be altered to appear like that of a complete stranger. When writing a paper, any researcher would want to provide readers with pictures as clear and easy to understand as possible. If one picture is unclear, the whole experiment should be repeated and a photo should be retaken. Yet in some cases, researchers instead amended the original pictures. It may have begun with a casual thought that it would be allowable to adjust contrast, but such adjustments could escalate infinitely, leading eventually to complete fabrication. Those who do not possess self-restraint, and thus cannot avoid this kind of temptation, do not qualify as researchers. Although the likelihood of problems occurring depends on the nature of individual study areas, experiments that nobody can reproduce are either fabrications or falsifications. Depending on the area of research, how easily a third-party can conduct a replication varies significantly. Researchers in areas where anybody in the world could conduct a replication within a few days after a paper is published cannot rely on fabrication or falsification. In contrast, animal tests for drug development are far more time consuming and troublesome to replicate, as one has to start by securing animals for experiments. Therefore, it would be impossible to conclude whether or not any false element was involved in the original tests. Wrongful acts tend to occur in this kind of area.

Dishonesty in clinical tests is sometimes reported by the mass media. As mentioned in Chapter Twelve, drug development takes time and the costs are considerable. Even if the basic research (drug discovery and exploratory research) is done and non-clinical trials are successful, the developer cannot recover their expenses unless the drug passes the clinical trials to be sold in the market. For pharmaceutical companies, the moment that determines whether a drug, which required so much time and money to develop, can be sold or not is critical. They can expect to achieve sales reaching tens of billions of yen per year if the drug passes the clinical trials. This motivation accounts for the misdeeds. For whatever reason, modifying or fabricating experimental data is not permissible. It could be a tragedy for a patient if a drug produced based on false data was not only ineffective, but also entailed hidden harmful side effects. Such cheating would be tantamount to a crime, rather than an ethical problem. Researchers and managers engaged in drug development are required to have a strict code of ethics. Needless to say, the same applies to the medical professionals who execute clinical trials. In terms of risk management regarding drug development, we need to connect multiple disciplines, such as public administration and the study of ethics, let alone pharmaceutical science, so that we can reach optimal solutions.

In the context of severe competition in scientific research, there is psychological pressure on researchers to produce positive results before others, and such pressure leads to misdeeds. Nevertheless, fabrication and falsification render precious time and money for scientific research futile, and eventually disrupt scientific progress. They may even cause societal distrust in scientists. In order for science to regain trust from society, each scientist has to become conscious of their responsibility, and relevant entities, such as universities and companies, need to seriously design and execute preventive measures against wrongful research.

Researchers' lack of ethics caused by the drastic slide towards fragmentation in disciplines such as science and engineering forms the backdrop of the current situation, in which risk management regarding scientific/engineering ethics is required. When we try to solve problems related to risk management in this area, the transdisciplinary approach of HSS will become increasingly indispensable.

Part V
Human Survivability Studies and Exploring the Future

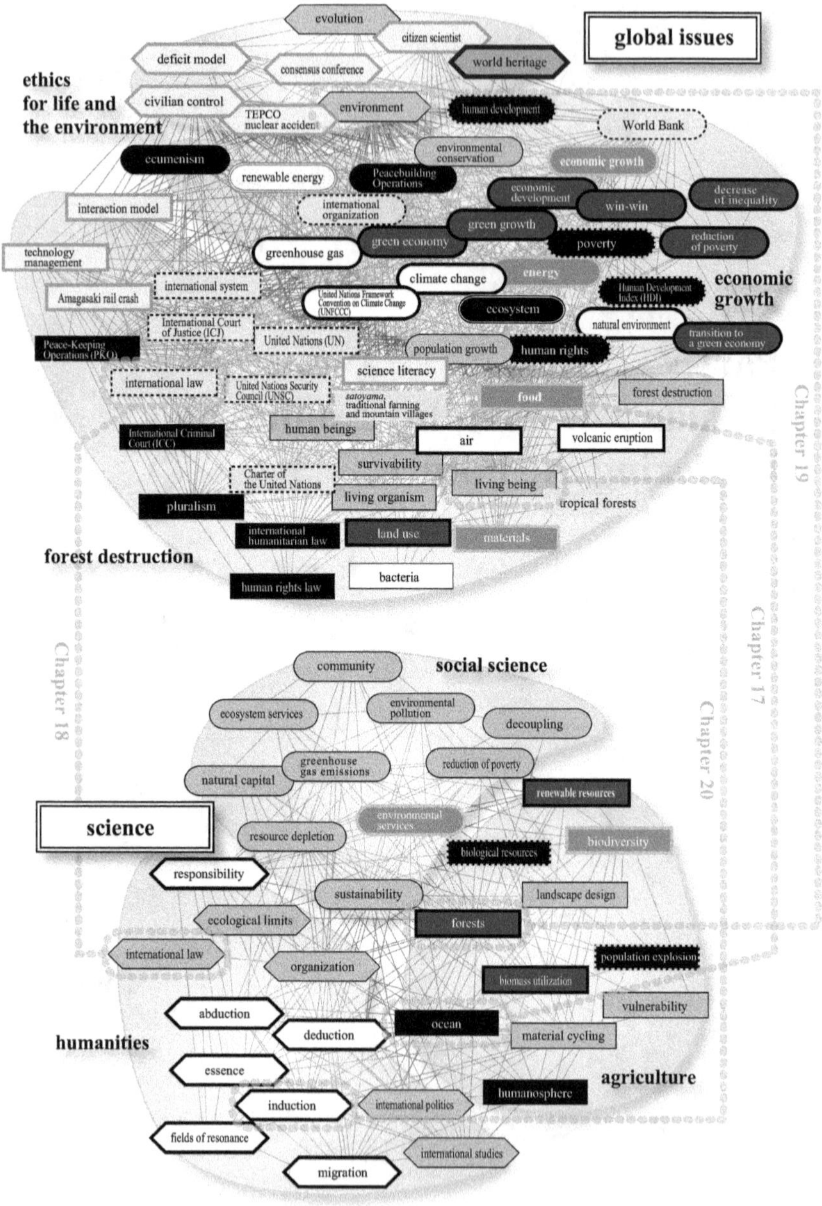

global issues
ethics for life and the environment
evolution
citizen scientist
deficit model
consensus conference
world heritage
civilian control
TEPCO nuclear accident
environment
human development
World Bank
environmental conservation
economic growth
ecumenism
renewable energy
Peacebuilding Operations
economic development
win-win
decrease of inequality
interaction model
international organization
green growth
technology management
green economy
greenhouse gas
poverty
reduction of poverty
climate change
energy
international system
Amagasaki rail crash
United Nations Framework Convention on Climate Change (UNFCCC)
Human Development Index (HDI)
economic growth
International Court of Justice (ICJ)
ecosystem
natural environment
United Nations (UN)
transition to a green economy
Peace-Keeping Operations (PKO)
population growth
human rights
science literacy
international law
United Nations Security Council (UNSC)
satoyama, traditional farming and mountain villages
food
forest destruction
International Criminal Court (ICC)
human beings
air
volcanic eruption
survivability
Charter of the United Nations
living being
living organism
tropical forests
pluralism
international humanitarian law
land use
materials
forest destruction
bacteria
human rights law
Chapter 19
Chapter 18
Chapter 17
Chapter 20
community
social science
ecosystem services
environmental pollution
decoupling
greenhouse gas emissions
reduction of poverty
natural capital
renewable resources
science
environmental services
resource depletion
biodiversity
biological resources
responsibility
sustainability
landscape design
ecological limits
forests
international law
population explosion
organization
biomass utilization
vulnerability
abduction
ocean
deduction
material cycling
humanities
essence
agriculture
humanosphere
induction
international politics
fields of resonance
international studies
migration

Introduction to Part V

In Part V, we raise a number of issues relevant to Human Survivability Studies (HSS) in relation to the future of humanity and the Earth system, with a view to achieving sustainable social and economic growth as well as rich and peaceful lives. In particular, we present new concepts, methodologies and practices concerning the environment, disputes, the economy and academia, and attempt to build a foundation for HSS.

In Chapter Seventeen, we consider ideas, lifestyles and other factors needed to build relationships of symbiosis and coexistence with nature. Focusing on forests, we consider ways to extricate ourselves from antagonism against and competition with nature and ecosystems, as foundations for human survival, and look at restoring a relationship of coexistence. We also examine conservation measures in this regard.

In Chapter Eighteen, we state that idealism and fundamentalism should be discarded in order to resolve disputes and conflicts between individuals, countries and regions and to achieve peace. We argue that we should confront the facts of disputes head on and search for their causes, devise and formulate systems for coordinating and resolving them and search for measures to effectively implement resolutions. We also explore the direction we should take in order to establish lasting peace.

Chapter Nineteen discusses green growth and green economics, i.e. economic development that vastly reduces environmental risk and ecological deficiency while achieving qualitative improvements to life and social equity. We also explore the possibility of a new framework for economic development that encourages a shift to a sustainable society.

In Chapter Twenty, we state the importance of abductive reasoning as a scientific method. We refer to specific cases in stating that trans-science, which deals with points of contact between domains of science and domains involved in policy and decision-making in politics and society, is indispensable for resolving social issues on a global scale. We show that it is important to produce HSS research that attempts to resolve issues from a holistic viewpoint, through broad 'transdisciplinary knowledge' and 'migration' across the humanities and sciences.

17 Coexistence with Nature

Shuichi Kawai

Protection and development

Ever since the human race came into being, humanity has coexisted with nature and all its many blessings. For example, we breathe air and use it to replenish water in order to live. Since we cannot convert inorganic to organic matter through photosynthesis, as plants do, we use other animals and plants as sources of nutrition. We have always coexisted with nature, and we have been blessed with life amid nature's material cycles and biological food chains. The general theory is that human history started in Africa and spread from there to other continents. This history can be described as one of opening up new frontiers and adapting to new local climates (Diamond 2000).

Today, however, hardly any primeval or 'unspoilt' nature remains in any terrestrial place capable of human habitation. Humans have modified nature to their own ends, produced food and created various artifacts. Humans have produced food via agriculture and livestock farming, mined for ores and quarried and processed various metals and fossil resources from nature. In the process, the relationship between humans and nature has not been one of coexistence, but rather one of antagonism. Specifically, agricultural and industrial revolutions have provided catalysts for the human race to flourish, but could also be said to have caused a widening rift between humans and nature. For example, the development of farmland through crop farming (cultivation of plants) and nomadism (domestication of animals) in the agricultural revolution destroyed or diminished forests and drove many locally unique species to extinction. The large-scale plundering of fossil resources since the industrial revolution has caused air pollution and climate change, while mining for ores and other underground resources has polluted our soil and water. The explosive growth of the human population in recent years has elevated the challenge of securing food, resources and energy to the greatest problem facing human survival. These concerns are in fact exacerbating our antagonistic relationship with nature.

Nineteen sixty-one is etched in history as the year when a human being embarked on the first manned flight in space. On that occasion, Yuri Gagarin of the then Soviet Union boarded Vostok 1 and orbited Earth. When he saw Earth covered by its beautiful blend of oceans, land and atmosphere (cloud), Gagarin uttered the now famous words 'The Earth is blue'. It was the moment when, for the first time in human history, humanity had left the Earth and looked down on it from above. At the same time, this event prompted the realization only too clearly that the Earth looks like a living entity, that the Earth is limited, or in other words, that it is 'Spaceship Earth' (see Chapter Two).

The crisis of human survival outlined in *The Limits to Growth* (1972), a report commissioned by the Club of Rome, was explored in greater detail and examined both comprehensively and analytically in the US government report *The Global 2000 Report* (1980). This trend was then continued in a series of UN conferences and others on the environment and development. In this way, concern over problems caused by the global environment grew significantly in the developed world from the second half of the twentieth century onwards. In the developing world, however, home to the majority of the world's population, the desire for economic growth in pursuit of lifestyles on par with those in developed nations makes it extremely difficult, in reality, to maintain a balance between 'protection and development'.

While it is self-evident that coexistence with nature is the key to human survival, the existence of the human race itself is turning into a kind of endlessly multiplying 'cancer' within the Earth's ecosystems. What we need at both an individual and social level are measures to control self-propagation, foster a spiritual culture and lifestyle that focuses on 'satisfaction with one's lot' and manage the desire for material possessions. We need to rethink how we can achieve sustainable use and balanced distribution of the Earth's limited resources, particularly bioresources and other renewables of a cyclical nature.

The challenge for Human Survivability Studies taken up in this chapter is that of coexistence between humans and nature. To sum this up in one phrase, we could frame it as 'retrieving nature'. While this entails extricating ourselves from antagonism and competition with nature and restoring a relationship of coexistence, specific measures to achieve this are difficult to formulate. Here, I consider how to ensure the sustainability of nature and ecosystems, which will be essential for human survival into the future, and discuss measures to bring this about. In particular, I introduce cases involving forests, as hubs of material cycles and ecosystems in the terrestrial region, and consider how to harmonize 'protection and development' through management aimed at securing resources while actively taking steps to regenerate degraded nature.

What is nature?

Of the *objects* in the external world surrounding ourselves as *subjects*, we use the term 'nature' to define a primeval world barely touched by human activity. The concept of 'nature' essentially refers to 'that which exists by itself' without human intervention – mountains, rivers, seas and the myriad things that live on and in them, as well as trees, plants and flowers, collectively known by some as 'the whole of creation'.

Domains of physical space involved in human survival and activity are described in their totality as the 'humanosphere'. Elements that comprise this humanosphere are broadly divided into three: the geosphere, the biosphere and the anthroposphere. The geosphere consists of the air surrounding the Earth (the atmosphere and outer space), the water and soil near its surface (hydrosphere and soilsphere) and the lithosphere beneath it, among others. This is the domain of the natural environment, comprising the world of material cycle networks. The biosphere, meanwhile, houses the loci of the life activity of living things, including plants and animals, but also fungi, bacteria and other microorganisms. This is seen as the world of ecosystem networks, or in other words, the eco-environment. Human beings are one of the species that exist in the biosphere. However, activity associated with subjective production and consumption by humans is now having a major impact on the geosphere and biosphere, and this interaction can no longer be ignored. For that reason, it would be more appropriate to treat the anthroposphere (specifically the sphere of human habitation) as a separate sphere. The anthroposphere is a domain of physical space consisting of human-made structures such as cities, houses, factories, roads and crop fields that lie at the heart of human activity. Based on the spheres described above and the interactive relationships between them, the concept of 'nature' could be understood as belonging to the biosphere and geosphere, the latter two of the environmental elements surrounding human beings. Sometimes these three spheres merge together to form and share spaces through mutual interaction.

Perceptions of the nature/human relationship

Primitive humans had an animistic view of natural phenomena, communing with nature as an object of awe, as something to be feared and worshipped. Nevertheless, in the epic poem *Gilgamesh* written in the ancient civilization of Sumer in Mesopotamian antiquity, nature is already perceived as something hostile to man; the dominant view is a dichotomy between man and nature which must be overcome and subjugated (see Chapter Four). Forests were a typical example of the latter, leading to a repeated process of tree felling and deforestation. In ancient cities, forests were destroyed to make way for the expansion of cities and farmland to cope with population growth. Large quantities of timber are also known to have been used in civil engineering works and buildings. This attitude to nature was maintained in ancient Greece, where nature was regarded as an opposing concept to human activity under the term *physis* (nature). Unlike these views that prevailed in Western Asia and Europe, in East Asia a binomially coexistential view of nature, in which humanity is harmoniously regarded as part of nature, has been widely accepted. Thus, the human attitude towards nature strongly reflects the climate, culture, history and religion of the region in question. If we take a general view of human and civilizational history, human beings could be said to have consistently destroyed nature and plundered food, timber, coal, oil and other resources and sources of energy from it. Land use has been utterly transformed; the expansion of agricultural practices converted forests into farmland and pastures, and these in turn into cities, factories and other domains alongside the growth of industry.

It is only in relatively recent times that terms like 'coexistence with nature' or 'symbiosis' have become commonplace. Until the modern era, nature was regarded as unfathomable, sometimes hostile, causing earthquakes, typhoons, floods, volcanic eruptions, forest fires and other natural disasters, as fearsome events that could not be managed or predicted by humans. Today, data from Earth observation satellites and others that offer a bird's eye view of Earth's systems are progressively being accumulated and analyzed. For example, the mechanism whereby solar energy concentrated in the tropics creates circulation in the atmosphere and oceans and is then transported to temperate regions where it takes the form of localized climate change and extreme weather, is gradually being unraveled. Similarly, the process whereby typhoons, cyclones and hurricanes spawned in the tropics absorb thermal energy and water vapor from the oceans and develop while moving toward temperate regions can also be predicted with considerable accuracy. These developments are gradually shaping our understanding of the effects caused by water, energy and other global material cycles on our weather via the atmosphere and oceans (see Chapter Eight). Meanwhile, the world of living organisms is adapting to localized weather and climate to build diverse ecosystems. The food chain is a causal relationship between organisms that is relatively easy to understand. In this way, nature depends on interactions between physical matter and living things.

Ways of achieving conservation, i.e. maintaining a relationship of symbiotic coexistence in which we sustain human life while preserving nature, have been the subject of much discussion in recent years. International conferences on 'environment and development' have been hosted about once every ten years by the UN and other peak bodies. The first 'United Nations Conference on the Human Environment' (Stockholm Conference) held in 1972 was followed by the Nairobi Conference in 1982, the Rio de Janeiro Conference in 1992, the Johannesburg Summit in 2002 and most recently Rio+20 (2012). At the Stockholm Conference, a broad-ranging discussion was held from the viewpoint of the anthroposphere, covering subjects such as human habitation, development and the environment, the holistic management of natural resources, humanity and the biosphere and environment-related information, culture and education. At the Rio Conference, agreement was reached in the form of the 'Rio Declaration on Environment and Development' as well as action plans for the Declaration such as 'Agenda 21' and the 'Statement of Forest Principles'. Also at Rio, frameworks for today's protection of the global environment and sustainable development were formed in the

'Framework Convention on Climate Change' and the 'Convention on Biological Diversity', among others.

Over these past twenty years, the limited nature of energy and resources as well as other 'limits of the Earth' have become increasingly clear, while the BRIC nations and other emerging economies have achieved standout economic growth. Experiences of disasters including the Sumatra earthquake and Indian Ocean tsunami and the Great East Japan Earthquake have led to a deeper awareness of risk management related to development and disasters. In the international community, there have been calls for a shift to a 'Green Economy', which seeks a balance between environmental conservation and economic growth (see Chapter Nineteen).

Why are forests important?

The most significant characteristic of living organisms is that they are 'reproductive' and that they maintain continuous cycles. On land inhabited by humans, forests are the most important components of the ecosystem. Although the Earth's forested area has decreased from six billion to four billion ha since the beginning of arable farming, it still covers one third of the world's total land area. Forests also retain an overwhelmingly large volume of reproducible biomass. The total volume of the Earth's biomass has been estimated at 1.8 trillion tons, of which accumulated forest biomass amounts to 1.65 trillion tons, thus accounting for more than 90% of the total (Sasaki 2007). On top of that, forests also play an important role as sites of interaction between ecosystems, material cycles, resource production and other components of the humanosphere.

Plants link inorganic with organic matter through water, air (oxygen) and carbon, and form the basis of natural and ecological networks that support a wide range of other organisms through primary production based on photosynthesis. As the amalgamation of woody plants, forests are hubs of terrestrial ecosystems and food chains, and together with the oceans, they are also hubs of material cycles for water, air, carbon and other elements in the geosphere. Forests are the most important places for the protection of terrestrial and brackish water ecosystems in the biosphere and provide a foundation for maintaining biodiversity and photochemical energy through photosynthesis. In terms of the anthroposphere, meanwhile, forests provide timber, food, medicines, fuel and other materials or energy sources. They are also sources of the water and oxygen essential for our survival. In this way, forests have various beneficial functions. The Science Council of Japan has estimated that the value of public benefit functions of Japan's forests amounts to around seventy trillion yen per year.

According to the most recent data by the United Nations Food and Agriculture Organization (FAO), the world's forests are shrinking by five million hectares per year (2000–2010; FAO 2011). This shrinkage is particularly conspicuous in South America, Africa, Southeast Asia and other regions where originally there were large accumulations of biomass and where there are tropical rainforests with the richest biodiversity. Of all Southeast Asian countries, the decline is particularly acute in Indonesia. According to the FAO, forested areas currently occupy 47% of the tropical zone sandwiched between the Tropics of Cancer and Capricorn on either side of the equator, while temperate forests account for 11% and boreal forests in high latitude regions account for 33%. The average biomass accumulations of these forests, i.e. the average existing volumes (volumes by dry weight), are said to be around 350–400, 300–350 and 200 t/ha (Sasaki 2007; Hara 2010), respectively. The annual biomass production of tropical forests, temperate forests and boreal forests (net ecosystem production) is around sixteen to twenty-two, twelve to thirteen and eight t/ha/y, respectively, showing that tropical forests are exceptionally voluminous in both forest accumulation and biomass production. These values are far larger than the average values for accumulation and biomass production of cropland (eleven t/ha and 6.5 t/ha/y, respectively), and are of course larger than the biomass accumulation (four to twenty-seven t/ha) and production (2.8 t/ha/y) of grassland and open woodland (Sasaki 2007).

Thus, tropical and subtropical natural forests, which account for the majority of the world's total forested area, are the largest existing sources of biomass accumulation and production. As stated above, the critical situation concerning the depletion of forests has sparked a thriving international debate on environment and development since the second half of the twentieth century. However, the discussion on 'sustainable development' tends to place more weight on the second of these elements, i.e. 'development'. As development in the sense of human beings modifying nature to suit their own objectives is essentially devoid of affinity with sustainability and maintenance, we must be careful of the environment being used as a cover for development.

Japan's forests and wood culture

The state of forests in Japan is very different from the global situation described above. Lying on the eastern edge of the Asia monsoon region, Japan's land mass stretches across a considerable expanse from north to south, is located in a warm and wet climate zone and has an environment eminently suited to plant growth. Japan is therefore blessed with rich forests and vegetation, mainly featuring temperate forests but also ranging from boreal to subtropical forests. The 'Age of the Gods' section of the ancient *Nihon Shoki* (Chronicles of Japan) reveals that the characteristics of representative trees like Japanese cedar, Japanese cypress and camphor were already well known in the ancient era, and that they were skillfully used for different purposes. In Japan, some 4,178 architectural structures are currently designated as national treasures or important cultural properties, and some of them have been registered as UNESCO World Cultural Heritage due to their significant historical and cultural value. More than 90% of these structures designated as cultural heritage are made of wood. A prime example is Hōryū-ji Temple in Nara, which has a history stretching back 1,300 years. This 'wood culture' is one outstanding characteristic of Japan's culture compared to that of western countries and even countries in East Asia and Southeast Asia, to say nothing of Western Asia.

In fact, Japan's forests suffered a period of depletion throughout the medieval Sengoku and Momoyama periods, when large quantities of timber were used for mass construction of shrines and temples, castles and other buildings. The forests recovered, however, thanks to plantation and strict regulatory policies by the feudal domains from the beginning of the Edo period in 1603 (Totman 1998). Compared to other parts of the world, where virgin forests were felled and then left to regenerate naturally, the Japanese developed intensive forest management technology including planting, weeding, pruning and thinning from extremely early on (second half of the sixteenth century), thus achieving a conversion from exploitative to sustainable forestry. A forest of Japanese cedar planted at the beginning of the Edo period can still be seen in Yoshino, Nara Prefecture (Tani

2008). These trees have grown to a breast-height diameter of about 1.5 m and a height of more than fifty m, creating the atmosphere of a virgin forest. In Germany and other parts of Europe, by contrast, forests were significantly devastated in the sixteenth and seventeenth centuries, when they were opened up for major expansion and conversion to farmland. Extensive (quasi-natural) forestry based on the principle of natural regeneration was promoted from the second half of the eighteenth century in order to recover and revive woodland, and steps were taken toward (environmental) protection and sustainable use of resources.

Once again greatly depleted by indiscriminate felling during World War II, Japan's forests recovered to a current area of 67% thanks to organized tree planting after the war, making Japan one of the most forest-rich countries in the developed world, along with Finland (74%). These human-made forests of Japanese cedar and cypress have grown to the point where timber can be harvested, but there has been little progress in thinning or logging, and Japan's self-sufficiency in timber remains low at 28%. As a result, domestic forestry has stagnated, the failure to make adequate use of trees has conversely led to a progressive disintegration of forests and there are fears over a decline in their environmental functions. For this reason, structural reforms of forestry and the development of infrastructure that can address changes in demand are pressing issues at the moment. The Forestry Agency has drawn up a Forests and Forestry Revitalization Plan (2009), and, reflecting this, a Forests and Forestry Basic Act (2011). These propose a number of systematic measures, starting with (1) achieving stable timber production by developing road networks, intensifying forest management and training foresters and management bodies, etc., and including (2) reforming the structure of domestic wood processing and distribution and (3) expanding the use of timber but also building networks directly linked to citizens and consumers, such as by using NPO activities and citizens' movements, and creating new types of forest industry and forest business (Forestry Agency 2001).

Attempts to regenerate forests

Regenerating forests by preventing deforestation and promoting afforestation of denuded land have become matters of pressing urgency in recent years. Although the world's human-made forested area is now gradually increasing at a rate of about 2.8 million ha per year (2000–2005), this still only accounts for 3.8% of the Earth's forests (140 million ha). The key to forest regeneration lies in aggressive afforestation, particularly on an industrial scale. This is because increasing manageable economic forests will provide direct incentives for forest regeneration, making it possible to furnish the increased demand for timber resources and energy accompanying population growth. This also constitutes a practical step toward maintaining existing virgin (reserved) and conservation forests.

Trees grow slowly when very young, then more vigorously, until growth slows again when they are mature. This results in a sigmoidal growth curve. However, the time span of the growth curve differs greatly depending on the region, species and management method, among other factors. For example, the endemic species known as Japanese cedar (*Cryptomeria japonica*) can maintain growth for 100 years or more with appropriate silvicultural management, while fast-growing tropical trees like the acacia (*Acacia spp.*) generally grow rapidly, but mature in ten to twenty years and then often stop growing (Kamis and Taylor 1993).

Whether in natural or man-made forests, biomass accumulation in stands is greatest at the mature stage. In mature forests, however, carbon absorption due to growth is balanced out by carbon emissions due to withering and decomposition, and the increase in biomass accumulation is thus zero. In the case of reserved forests, where the priority is on biodiversity maintenance functions, biomass accumulation is important as an evaluation index. In economic forests (production forests) that prioritize timber production functions, on the other hand, increasing stock is important. In production forests, trees are felled to match the increase in stands, then processed and used to the maximum possible effect, thus ensuring continuity of production and use. To this end, it is vital that annual biomass accumulation (stock), the increase in stock and logging (flow) are analyzed and understood dynamically.

Unregulated development and excessive felling have decimated forests and turned them into grassland in southern Sumatra, Indonesia. Figure 17.1 compares the biomass stock and flow in natural forests, grassland and acacia forests in that region (as of 2006) based on a case of large-scale industrial afforestation that started in 1990 (Kawai 2016; Kobayashi et al. 2015). In the figure, the biomass accumulation of tropical rainforests is largest with a maximum 400

Figure 17.1: Biomass accumulation (stock) and logging (flow) in natural forests, plantations and grassland

Source: Kawai (2016).

t/ha, but these are mature forests, where carbon absorption due to growth is balanced out by carbon emissions due to withering and decomposition. As a result, the increase in carbon accumulation is evaluated as zero. Moreover, if there were no removal of wood from the forest due to illegal logging or similar, the flow would again be zero. In other words, biomass in tropical natural forests shows a state of apparent stability. In the same way, biomass accumulation on grassland dominated by cogon grass (*Imperata cylindrica*) is only around four t/ha. Furthermore, much of the biomass growth corresponding to the annual increase in stock inside the forest is lost to withering and decomposition, and the flow discharged outside the system due to energy, etc., is estimated to be at most around 70% of the total (2.8 t/ha/y). By contrast, in acacia plantation areas aged one to six years, a cycle of felling and plantation is repeated over a certain area every year. Thus, while there is variation in the accumulation, increase and logging volumes in individual stands, a certain stable volume of accumulation and increase may be expected over the afforested area as a whole (120,000 ha), while logging is carried out every year. That is, the figure shows that as of 2006, the average accumulation volume of one to five year old acacia forests was seventy-eight t/ha, which is not necessarily large compared to that of tropical rainforests. However, when the logging flow of around sixty-two t/ha/y is added, the total of this stock and flow in 2006 is estimated to be 139 t/ha. This could be seen as significant in terms of forest conservation efforts.

Opening up the future: Coexistence with nature

What sort of landscape lies behind our visualization of coexistence with nature?

For example, the Japanese concept of *satoyama*, a scene of traditional farming and mountain villages, depicts a landscape created through interaction between humans and nature. Primeval nature comprising forests of evergreen, deciduous broad-leaved and other shade trees (tolerant trees) has been destroyed by anthropogenic disturbance, causing a gradual loss of fertility. In its place, forests of Japanese oak, Japanese red pine, chestnut and other sun trees (intolerant trees) grow and become intermixed, forming what is known as mixed woodland. Local people practice a repeated cycle of appropriate felling in mixed woodland around villages, gather fallen leaves and branches as fuel or fertilizer, and thereby maintain a stable ecosystem, not in a state of climax but in a state of transition from deciduous forests. This is what we call *satoyama*. 'Transition' is the process where living organisms modify the environment they have adapted to through their own presence and prosperity, where other species inhabiting that space also change. In other words, *satoyama* could be said to refer to secondary forests modified by human activity, or managed nature. The *satoyama* landscape, consisting of villages and the countryside (paddy fields) around them, along with the mixed woodland and plantation forests of Japanese cedar, cypress and others behind them, could be described as a classic Japanese *furusato* (homeland) scene. This *satoyama* landscape in which humans coexist with nature can similarly be observed in places all over the world.

Besides transitions in ecosystems caused by climate change, the interaction between living organisms and the environment, whereby humans and living organisms have an effect on the environment and the environment changes correspondingly while living organisms also adapt to this, is widely known through research on transitions in vegetation based on pollen analysis, radiocarbon dating and others (see Chapter Four). The *satoyama* landscape has been created through a system designed to support paddy rice cultivation since the Yayoi period – in other words, a system of sustainable and stable food, energy and material production – and has played a major role in forming Japan's rural landscape and *satoyama* culture. *Satoyama* and *satoyama* culture used to support the lives of the Japanese, and moreover came with conditions that were maintained over the long-term. In other words, they were established as a model of coexistence between humans and nature. At the same time, they have played a major role in forming the Japanese view of nature, in which *sato* (human society), *satoyama* and remote mountains (nature) are viewed as a continuum.

Satoyama could also be seen as a buffer zone between nature and humans. In eras when anthropogenic disturbance is more pronounced, the *satoyama* environment is devastated and mixed woodland cannot be maintained, transitioning to grassland and pine forests. This tendency is particularly strong in suburban *satoyama*, where anthropogenic disturbance is more conspicuous. Archive records like *Miyako Meisho Zue* (Images of famous places in the capital, Kyoto) suggest that *satoyama* in the suburbs of Kyoto were severely disturbed by humans collecting timber, fertilizer and fuel (firewood) in the Edo period. The ecosystems fell into a state of undernourishment, or an extremely depleted state featuring many bare mountains with only a few pines dotted sporadically. However, when people stop collecting firewood and fertilizer from *satoyama*, as in recent years, those *satoyama* become rich in nutrition and transition to forests dominated by laurel. At the same time, deer, boars, monkeys and other wild animals have proliferated, and together with a shortage of nuts, mushrooms and other natural food sources, they appear in villages and towns and create problems there. In this way, *satoyama* are also places where humanity's relationship with nature appears most conspicuously. While they are symbolic of the sustainable use of nature, it is also evident that their maintenance and conservation depends on appropriate management.

As stated above, if we look back over the history of humanity's relationship with nature, it has become antagonistic where only the human side profits. This should be corrected to the original symbiotic relationship where both sides benefit. Although it is of course important to preserve primeval nature as it is, it is regrettable that nature in a primeval state hardly exists anywhere today. Measures for conservation, whereby preservation and exploitation are balanced through appropriate management of nature, are important in practice (see Chapters Three and Four).

Forests may be increased through management, and in this respect they differ greatly from mineral and fossil resources. On the other hand, forests can rapidly decline and be destroyed or devastated by excessive felling and inadequate management. The recent decline of tropical rainforests in South America and Southeast Asia is often highlighted. In the tropics, once land becomes bare, runoff of soil nutrients is vigorous and vegetation is extremely slow to recover. To regenerate forests in tropical regions, therefore, renewing plantations will be more effective than regenerating natural forests, and regenerative forestry should be promoted for the conservation management of tropical forests. Besides technical issues that need to be resolved for industrial afforestation in the tropics, however, there are still many environmental and social issues. For example, the degree to which short rotation forestry causes loss of nutrients and soil deterioration is unknown. Besides a decline in biodiversity and increased vulnerability to disease and pests due to large-scale afforestation, other issues include friction with local inhabitants. Instead of the clear-felling method that has a significant impact on nature, attempts at forest management in which locally dominant species are planted in the spaces left after line thinning of natural secondary forests have been started on the Indonesian island of Kalimantan (Borneo) and elsewhere. In this way, there is an ongoing search for 'coexistence with nature' in which a compromise is sought between economic growth and environmental issues.

In Japan, the importance lies more in making positive use of timber resources and caring for human-made forests and mixed woodland in *satoyama* for the sake of environmental conservation. This is because conserving soil and water through afforestation helps us to make full use of the role of biodiverse forests as environmental resources. The nature of river basins, ranging from remote mountains to *satoyama, sato* (human habitation), *satogawa* (managed rivers) and on to *satoumi* (managed seas), in combination with human activity, is a classic example of a sustainable society that is rich in harmony. *Sato*, i.e. humans, link forests with seas and enrich nature (Mukai 2012). Sustainable use of *satoyama* resources, as loci for local development, is expected to provide a foothold for community regeneration, rich lifestyles and the preservation of traditional culture. To forge a future life in which nature is used sustainably with coexistence between humans and nature, we need to learn from history and the wisdom of our forebears, reappraise multifaceted values and search for new lifestyles, values and social development. We need to shift from a doctrine of growth and expansion to one of coexistence and symbiosis that emphasizes sustainable cycles.

From our consideration of forests, we need to ensure the continuity of natural environments and ecosystems as well as the sustainability of timber production. To this end, regions will need to be zoned over wide areas in accordance with multifaceted functions. This will require landscape design that aims to harmonize the anthroposphere with the natural sphere. In terms of the biosphere, we need landscape design that appropriately allocates reserved forests that focus on ecosystem maintenance and conservation functions, secondary forests (conservation forests) as buffer zones and production forests that emphasize the function of supplying resources and energy in the anthroposphere. We will need to develop technology that seeks harmony with the environment as a foundation for human activity and survival, and to create doctrines and systems for a return to nature.

Research on the diversity of bird species in large-scale industrial afforestation in tropical regions, as seen above, shows that isolated and scattered small-scale secondary forests left in afforested areas provide important habitats for bird populations, and contribute to maintaining the diversity of species (Fujita et al. 2014). In acacia afforested areas with young trees (two years old), the diversity and habitat density of bird species are poor, with diversity close to that of bare land or grassland. But in acacia forests aged four or more years, diversity has recovered to a level approximating that of secondary forests. These research results suggest that afforested areas in which a certain level of bird species diversity can be secured are established by appropriately allocating secondary forests and appropriately designing forest stand demarcation, even in single-species large-scale industrial afforestation. For animals that move on the ground, however, it is unknown what role is played by isolated secondary forests. Our challenge will be to secure conservation forests that are linked together providing animal corridors. Pressing tasks in this regard will include grasping the current state of land use transition and forest resources from a bird's-eye view and with high precision, developing systems for holistic management of bioresources and achieving future landscape design for coexistence with nature, by harnessing the satellite remote sensing technology that has seen such conspicuous growth in recent years (Kobayashi et al. 2015).

To summarize, issues related to humanity's coexistence with nature bring into question our human doctrines and lifestyles, as well as the nature of society. Conceptualizing the nature of a harmonious human society, such as looking at the best ways of using biomass, developing local communities and correcting disparity and poverty, will also be important in terms of ensuring the sustainability of nature.

18 Resolving Conflicts and Achieving Peace

Makoto Ohishi

Introduction

Resolving conflicts

As mentioned in Part III (see Chapter Nine), where the relationship between Human Survivability Studies and the issues faced by modern society was discussed, contemporary society is confronting a wide range of conflicts at both the domestic and international level. Since it is not possible to discuss all aspects of these conflicts here, I have selected a few and will examine efforts that have been made towards their resolution.

One international movement focused on religious conflicts is Eucumenism, a movement toward the unity of different Christian traditions (also called 'world church movement', or 'world churches cooperation movement' in Japan). From this meaning, the term 'Eucumenism' may also be used to talk about the dialogue and cooperation between different churches outside the Christian faith. Nevertheless, it originally referred to the efforts of the World Council of Churches (WCC), a movement started by Protestant and Eastern Orthodox churches and joined by the Catholic Church after the Second Vatican Council (1962–65).

Another movement fostering dialogue between churches is Religions for Peace (WCRP). This movement was founded in 1961 during the most dramatic period of the Cold War, when, feeling the need to take action towards world peace, the leaders of different religions around the world held a World Summit of Religious Leaders. From 1970, it became known as Religions for Peace International, an international organization based on a coalition between the world's main religions that aims to promote joint efforts towards world peace and enjoys NGO status with the United Nations Economic and Social Council (ECOSOC).

The first World Conference was convened in Kyoto from October 16 to 21, 1970. The Japanese committee was created as a public interest incorporated foundation, with the stated purpose to 'promote a closer relationship based on the spirit of tolerance and deep reconsideration of the past of exclusion and dogmatism' and 'gather the knowledge that comes from interfaith dialogue and mutual understanding, and join efforts to achieve peace'. The membership of the committee is comprised of the Japan Buddhist Federation, the Jinja Honcho (Association of Shinto Shrines), the Japan Confederation of Christian Churches, the Kyouha Shinto (Sect Shinto) and the Federation of New Religious Organizations of Japan, which are members of the Japanese Association of Religious Organizations.

Recognition of pluralism

As discussed earlier in this volume (see the conclusion of Chapter Nine), the model that seeks the peaceful coexistence of a variety of cultural groups while maintaining their cultural identities, i.e. the pluralist model, is one of the fundamental frameworks for conflict resolution. In this case, the question is how to get each cultural group to accept pluralism through dialogue and mutual understanding.

In this context, expectations increase towards international organizations and entities, which operate across multiple ethnic, cultural, religious and national borders. In the following pages I outline these kinds of efforts, focusing on the work of the UN.

Basic global tasks

Main problems and principles of action of the UN

The United Nations deals with four main concerns, stated as its 'purposes' in article 1 of the UN Charter as follows: (1) maintaining international peace and security; (2) developing friendly relations among nations based on respect for the principle of equal rights and self-determination of peoples; (3) achieving international cooperation in solving international problems of an economic, social, cultural or humanitarian character, and in promoting and encouraging respect for human rights and for fundamental freedoms; and (4) to be a centre for harmonizing the actions of nations in the attainment of these common ends (see the United Nations Information Centre summary). The reason behind the first purpose of the UN is related to the motive that led to its establishment.

The principles of action adopted by the UN for achieving its purposes are also specified in the Charter (art. 2). In short, they are: (1) the UN is based on the principle of the sovereign equality of all its Members; (2) all Members shall fulfill in good faith the obligations in accordance with the UN Charter; (3) all Members shall settle their international disputes by peaceful means; (4) all Members shall refrain in their international relations from the threat or use of force against any state; (5) all Members shall give the UN every assistance in any action it takes in accordance with the UN Charter; and (6) the UN is not to intervene in matters which are within the domestic jurisdiction of any state[1].

Expansion of the UN's activities

According to the home page of the United Nations Information Centre, the United Nations, based on the above-mentioned goals and principles of action, is engaged in 'safeguarding peace, protecting human rights, protecting the environment, promoting democratization, fighting against terrorism, etc.', and locates these as global issues. Needless to say, this is only a list of the key current issues, and it does not follow that the UN's activities are limited to these.

In fact, at the beginning of the '60 Ways the United Nations Makes a Difference', published by the Information Centre, it is stated that the range of activities of the UN has expanded to:

> child survival and development, environmental protection, human rights, health and medical research, alleviation of poverty and economic development, agricultural development and fisheries, education, the advancement of women, emergency and disaster relief, air and sea travel, peaceful uses of atomic energy, workers' rights, etc.

This book is dealing with many of the above problems. Nevertheless, we need to keep in mind that in reality, there are numerous obstacles on the road to solving these issues.

If we take a look at the purpose of achieving international 'peace', we will notice that the United Nations Security Council, the organ that was expected to play a central role in achieving this purpose, has always been criticized for its ineffectiveness in maintaining or restoring peace in the international community due to the 'veto power' that the five permanent member countries are vested with. To be more precise, research has been conducted that shows that from the date of foundation of the UN up to 2012, 269 vetoes have been cast, of which nearly 244 were exercised during the Cold War period, and from 1996 to 2012 in the period after the Cold War, the sum of vetoes cast by the US, Russia and China was twenty-five[2].

Response of the UN system

In order to deal with the basic issues mentioned above, the United Nations Charter has established six principal organs, namely, the General Assembly, the Security Council, the Economic and Social Council, the Trusteeship Council, the International Court of Justice and the Secretariat (art. 7 of the Charter)[3].

The UN Charter also authorizes the General Assembly and the Security Council to establish subsidiary organs (art. 22 and 29), and allows the Economic and Social Council to enter into agreements with specialized agencies (agencies established by intergovernmental agreement and brought into relationship with the United Nations in accordance with the procedures stated in the Charter. See art. 57 and 63).

With regard to the subsidiary organs, the General Assembly has established various permanent committees, such as a disarmament committee and an international law committee, and also created twenty-four plans and funds as the United Nations Conference on Trade and Development (UNCTAD), the United Nations Environment Programme (UNEP), the United Nations Children's Fund (UNICEF) and the United Nations World Food Programme (WFP), etc. The Security Council has established the Counter-Terrorism Committee, the Military Staff Committee and the Political and Peacekeeping Missions, etc. On the other hand, the Economic and Social Council has entered into agreements with fifteen agencies such as the International Labour Organization (ILO), the Food and Agriculture Organization (FAO), the United Nations Educational, Scientific and Cultural Organization (UNESCO), the World Health Organization

(WHO), the International Monetary Fund (IMF) and the International Civil Aviation Organization (ICAO), etc. In addition, other organs related to the United Nations were also established, such as the International Atomic Energy Agency (IAEA), which is responsible for submitting reports to the General Assembly and the Security Council, and the World Trade Organization (WTO), which does not have to report to the General Assembly but contributes, in an ad hoc manner, to the discussions of the Economic and Social Council about finance and development.

In this way, as shown in Figure 18.1, the system of the UN is comprised of six principal organs and forty other agencies and organizations, making the whole of its structure complex and hard to understand. Nevertheless, it is through this very structure that the UN responds to the issues in which it is engaged.

International law and the rule of law

Establishment of international law

According to 'L'ABC des Nations Unies' of the UN's Department of Public Information, one of the most important achievements of the UN is that it has 'established a system of international law, which consists of treaties, agreements, standards, etc., that makes a decisive contribution towards economic and social development and the promotion of international security and peace'.

Today, it is said that more than 500 multilateral agreements have been signed under the UN, as agreements on environmental protection, regulation of migrant labor, policies against drug trafficking, counter-terrorism, etc. There are also many other treaties and agreements in the field of international human rights law in the form of the International Covenant on Civil and Political Rights (ICCPR), the International Covenant on Economic, Social and Cultural Rights (ICESCR), the International Convention on the Elimination of All Forms of Racial Discrimination (ICERD), the Convention on the Rights of the Child (UNCRC), etc., and in the field of international humanitarian law, the Convention on the Prevention and Punishment of the Crime of Genocide (CPPCG) and the Geneva Conventions on the protection of victims of war, etc.

There is no doubt that despite the fact that these international laws have been established, if there are no sanctions imposed in response

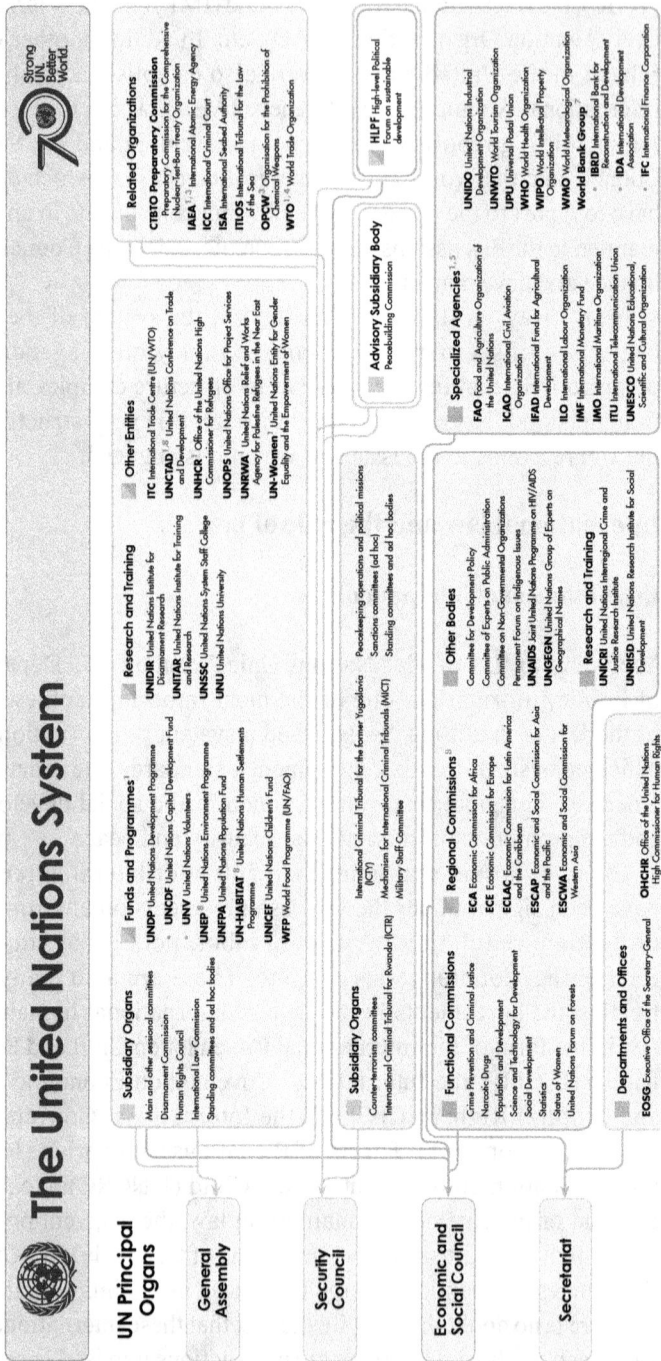

The United Nations System

70 Strong UN. Better World.

UN Principal Organs

General Assembly

Security Council

Economic and Social Council

Secretariat

Subsidiary Organs
- Main and other sessional committees
- Disarmament Commission
- Human Rights Council
- International Law Commission
- Standing committees and ad hoc bodies

Funds and Programmes[1]
- **UNDP** United Nations Development Programme
  - **UNCDF** United Nations Capital Development Fund
  - **UNV** United Nations Volunteers
- **UNEP[2]** United Nations Environment Programme
- **UNFPA** United Nations Population Fund
- **UN-HABITAT[3]** United Nations Human Settlements Programme
- **UNICEF** United Nations Children's Fund
- **WFP[8]** World Food Programme (UN/FAO)

Research and Training
- **UNIDIR** United Nations Institute for Disarmament Research
- **UNITAR** United Nations Institute for Training and Research
- **UNSSC** United Nations System Staff College
- **UNU** United Nations University

Other Entities
- **ITC** International Trade Centre (UN/WTO)
- **UNCTAD[4,8]** United Nations Conference on Trade and Development
- **UNHCR[5]** Office of the United Nations High Commissioner for Refugees
- **UNOPS** United Nations Office for Project Services
- **UNRWA[6]** United Nations Relief and Works Agency for Palestine Refugees in the Near East
- **UN-Women[7]** United Nations Entity for Gender Equality and the Empowerment of Women

Related Organizations
- **CTBTO Preparatory Commission** Preparatory Commission for the Comprehensive Nuclear-Test-Ban Treaty Organization
- **IAEA[1,3]** International Atomic Energy Agency
- **ICC** International Criminal Court
- **ISA** International Seabed Authority
- **ITLOS** International Tribunal for the Law of the Sea
- **OPCW[3]** Organization for the Prohibition of Chemical Weapons
- **WTO[1,4]** World Trade Organization

Subsidiary Organs
- Counter-terrorism committees
- International Criminal Tribunal for Rwanda (ICTR)
- International Criminal Tribunal for the former Yugoslavia (ICTY)
- Mechanism for International Criminal Tribunals (MICT)
- Military Staff Committee
- Peacekeeping operations and political missions
- Sanctions committees (ad hoc)
- Standing committees and ad hoc bodies

Advisory Subsidiary Body
- Peacebuilding Commission

HLPF High-level Political Forum on sustainable development

Functional Commissions[8]
- Crime Prevention and Criminal Justice
- Narcotic Drugs
- Population and Development
- Science and Technology for Development
- Social Development
- Statistics
- Status of Women
- United Nations Forum on Forests

Regional Commissions[8]
- **ECA** Economic Commission for Africa
- **ECE** Economic Commission for Europe
- **ECLAC** Economic Commission for Latin America and the Caribbean
- **ESCAP** Economic and Social Commission for Asia and the Pacific
- **ESCWA** Economic and Social Commission for Western Asia

Other Bodies
- Committee for Development Policy
- Committee of Experts on Public Administration
- Committee on Non-Governmental Organizations
- Permanent Forum on Indigenous Issues
- **UNAIDS** Joint United Nations Programme on HIV/AIDS
- **UNGEGN** United Nations Group of Experts on Geographical Names

Research and Training
- **UNICRI** United Nations Interregional Crime and Justice Research Institute
- **UNRISD** United Nations Research Institute for Social Development

Specialized Agencies[1,5]
- **FAO** Food and Agriculture Organization of the United Nations
- **ICAO** International Civil Aviation Organization
- **IFAD** International Fund for Agricultural Development
- **ILO** International Labour Organization
- **IMF** International Monetary Fund
- **IMO** International Maritime Organization
- **ITU** International Telecommunication Union
- **UNESCO** United Nations Educational, Scientific and Cultural Organization
- **UNIDO** United Nations Industrial Development Organization
- **UNWTO** World Tourism Organization
- **UPU** Universal Postal Union
- **WHO** World Health Organization
- **WIPO** World Intellectual Property Organization
- **WMO** World Meteorological Organization
- **World Bank Group**
  - **IBRD** International Bank for Reconstruction and Development
  - **IDA** International Development Association
  - **IFC** International Finance Corporation

Departments and Offices
- **EOSG** Executive Office of the Secretary-General
- **OHCHR** Office of the United Nations High Commissioner for Human Rights

International Court of Justice

Trusteeship Council[8]

DESA Department of Economic and Social Affairs
DFS Department of Field Support
DGACM Department for General Assembly and Conference Management
DM Department of Management
DPA Department of Political Affairs
DPI Department of Public Information
DPKO Department of Peacekeeping Operations
DSS Department of Safety and Security
OCHA Office for the Coordination of Humanitarian Affairs

OIOS Office of Internal Oversight Services
OLA Office of Legal Affairs
OSAA Office of the Special Adviser on Africa
PBSO Peacebuilding Support Office
SRSG/CAAC Office of the Special Representative of the Secretary-General for Children and Armed Conflict
SRSG/SVC Office of the Special Representative of the Secretary-General on Sexual Violence in Conflict
UNISDR United Nations Office for Disaster Risk Reduction

UNODA United Nations Office for Disarmament Affairs
UNODC[1] United Nations Office on Drugs and Crime
UNOG United Nations Office at Geneva
UN-OHRLLS Office of the High Representative for the Least Developed Countries, Landlocked Developing Countries and Small Island Developing States
UNON United Nations Office at Nairobi
UNOP[2] United Nations Office for Partnerships
UNOV United Nations Office at Vienna

Notes:
1 All members of the United Nations System Chief Executives Board for Coordination (CEB).
2 UN Office for Partnerships (UNOP) is the UN's focal point vis-à-vis the United Nations Foundation, Inc.
3 IAEA and OPCW report to the Security Council and the GA.
4 WTO has no reporting obligation to the GA, but contributes on an ad hoc basis to GA and Economic and Social Council (ECOSOC) work on, inter alia, finance and development issues.
5 Specialized agencies are autonomous organizations whose work is coordinated through ECOSOC (intergovernmental level) and CEB (inter-secretariat level).
6 The Trusteeship Council suspended operation on 1 November 1994, as on 1 October 1994 Palau, the last United Nations Trust Territory, became independent.
7 International Centre for Settlement of Investment Disputes (ICSID) and Multilateral Investment Guarantee Agency (MIGA) are not specialized agencies but are part of the World Bank Group in accordance with Articles 57 and 63 of the Charter.
8 The secretariats of these organs are part of the UN Secretariat.

This Chart is a reflection of the functional organization of the United Nations System and for informational purposes only. It does not include all offices or entities of the United Nations System.

Figure 18.1: The United Nations system
Source: www.un.org/en/aboulun/structure/pdfs/17-00023e_UN%620System%20Chart_8.5x11_4c_EN_web.pdf (UN system chart)

to violations or infringements, they would not be effective as laws. For this reason, some treaties include specific sanctions. Yet, it is desirable to establish permanent courts to deal with international disputes similar to those used by domestic institutions for resolving domestic disputes.

International Court of Justice (ICJ)

In this context, the UN Charter has established the International Court of Justice as the 'principal judicial organ of the United Nations' in Chapter XIV (art. 92), stating that all members of the United Nations are ipso facto parties to the 'Statute of the International Court of Justice', which is considered part of the UN Charter (art. 93). Therefore, each member state has the duty 'to comply with the decision of the International Court of Justice in any case to which it is a party', and if a member state fails to comply, the other party may have recourse to the Security Council, which may make the necessary recommendations or decide upon measures to be taken (art. 94).

What is important here is that while other principal organs of the UN are located in the US, only the ICJ is located in the Netherlands (art. 22 of the ICJ Statute). This indicates that the ICJ has followed the example of the Permanent International Court of Justice established by the League of Nations; we should also note that this location is symbolic of judicial independence. In addition, the ICJ consists of fifteen members elected by the General Assembly and the Security Council for nine-year terms, with the possibility of reelection (art. 3, 4 and 13 of the ICJ Statute).

The basic role of the ICJ is to decide disputes submitted to it in accordance with international law (judicial authority), applying international conventions, international custom and the general principles of law, etc. (art. 38). However, the ICJ also has the authority to give advisory opinion on legal questions at the request of specialized agencies or organs of the UN such as the General Assembly or the Security Council (art. 96 of the UN Charter and art. 65 of the ICJ Statute).

Either way, there is no doubt that the precedents accumulated by the ICJ play an important role in the field of international law. Nevertheless, the ICJ is a 'civil law court' for deciding legal disputes between countries, thus, it does not have jurisdiction to prosecute individuals, as in criminal jurisdiction.

International Criminal Court

For this reason, after the serious infringements of international humanitarian law that occurred during the genocides in Yugoslavia and Rwanda, the Security Council, based on Chapter VII of the UN Charter, established international criminal courts as subsidiary organs with jurisdiction limited to a specific purpose and region for prosecuting this sort of crime (the International Criminal Tribunal for the former Yugoslavia or ICTY was established in 1993, and the International Criminal Tribunal for Rwanda or ICTR was set up in 1994). In some cases, extraordinary criminal chambers are established to rule on the infringement of international humanitarian laws and other related legislation through agreements between the UN and concerned states (the Special Court for Sierra Leone in 2002, the Extraordinary Chambers in the Courts of Cambodia in 2006 and the Special Tribunal for Lebanon in 2009).

Nevertheless, these international criminal courts are just ad hoc tribunals, which close upon the achievement of their purpose. For this reason the International Criminal Court (ICC) was founded as a permanent court by the International Criminal Court Statute (also referred to as the Rome Statute), which was adopted in July 1998 and came into force in July 2002.

The ICC consists of eighteen judges elected for nine-year terms, who, as a general rule, are not eligible for re-election (art. 36 of the Rome Statute). It is a permanent and independent court located at The Hague, as is the International Court of Justice, with jurisdiction over serious crimes of concern to the international community, such as genocide, crimes against humanity and war crimes (art. 3, and 5 to 8 of the Statute). We should note, however, that although the ICC may cooperate with the UN, it is not an organ of the UN.

Peace and security

UN peacekeeping operations

As mentioned above, the primary purpose of the UN is 'maintaining international peace and security'. However, the organ designed to play a central role in achieving this goal, i.e. the Security Council, has been unable to perform this task and did not meet initial expectations of maintaining or restoring peace in the international community, mainly due to the difficulty of making its permanent members, which have veto power, cooperate during the period when the Cold War was escalating.

The above-mentioned UN Peacekeeping Operations (PKO), which has the consent of the parties, impartiality and non-use of force as its basic principles, emerged as an alternative organ to perform this task. It has a long history, which according to official UN papers, dates back to 1948. In Japan, however, it started to draw attention upon the enactment of the PKO Law[4] after the invasion of Kuwait by Iraqi forces in September 1990 and the subsequent Gulf War, triggered by the annexation of Kuwait (1991).

According to UN statistical data, a total of sixty-nine peacekeeping operations were conducted from 1948 to October 2014, and at the present moment (October 31, 2014), there are sixteen operations underway, including the United Nations-African Union Mission in Darfur, the United Nations Organization Stabilization Mission in the Democratic Republic of Congo (MONUSCO) and the United Nations Mission in South Sudan (all established in July 2007). Up to now, a total of 120,622 persons, mainly uniformed personnel, from more than 160 countries have participated in the operations, and a total of 3,308 persons (UN personnel) have lost their lives. We cannot forget what the UN has achieved thanks to these efforts: for example in sixty-seven UN missions, citizens of a few dozen countries were able to participate in free elections and more than 400,000 former soldiers were disarmed.

Note, however, that the UN needs a large amount of funds to continue their peacekeeping operations; the budget from July 2014 to June 2015 was 7.006 billion dollars (US). The peacekeeping operations budget is separated from the regular UN budget and is financed mainly by member states (the contribution rate between 2013 and 2015 was 28.39% from the US, 10.83% from Japan, 7.21% from France, 7.14% from Germany, 6.68% from England and 6.64% from China, etc.), but there is unfortunately no space to go into details here.

Peacebuilding operations

The role that the UN is supposed to play in order to achieve 'international peace and security' is not limited to the prevention of disputes and the maintenance and restoration of peace. Recently, the UN has begun to engage in so-called 'peacebuilding operations'.

'Peacebuilding operations' is a general term that refers to the breadth of activities aimed at building the basic structures for achieving sustainable peace by avoiding situations where a country enters into or reenters a state of conflict and strengthening its conflict management capacity. In 2006, the Peacebuilding Commission (PBC) was established as an advisory and subsidiary body of the UN with the purpose of helping the transition of concerned countries from a state of war to one of sustainable peace.

The most important actors in peacebuilding operations are civilians rather than military personnel (note, however, that in many cases military and police advisers join the operations). As of today, thirteen peacebuilding/restoration missions have been dispatched and currently, as shown in Figure 18.2, eleven political and peacebuilding missions are being conducted in, for example, West Africa, Somalia, the Middle East and Afghanistan (as of February 28, 2015).

Theories and policies on peace and security

We have examined efforts towards peace focusing mainly on the implementation of 'international peace and security' by the UN. Finally, I discuss the theories and policies that international relations and international political science have provided for dealing with the reality of international relations and politics.

In order to analyze the situation regarding the resolution and conciliation of international disputes, it is important to understand the theories and policies on international peace and security that are engaged in investigating the efficacy and effectiveness of peacebuilding measures based on the principal theories of international relations. The three basic schools of international relations are as follows (see Suzuki Motoshi 2007):

1. 'Realism' is the school of international relations theory that focuses on the probability that a state may use force as a real threat and aims to establish an 'uncertain peace' based on controlling the international balance of power through a system of 'checks and balances'.

2. 'Institutionalism' is the school of international relations theory that focuses on the risk of mutual distrust and uncertainty and aims to achieve stability in the form of a 'temporary peace' based on the management of this risk through the procedures and norms of international institutions.

UNITED NATIONS POLITICAL AND
PEACEBUILDING MISSIONS

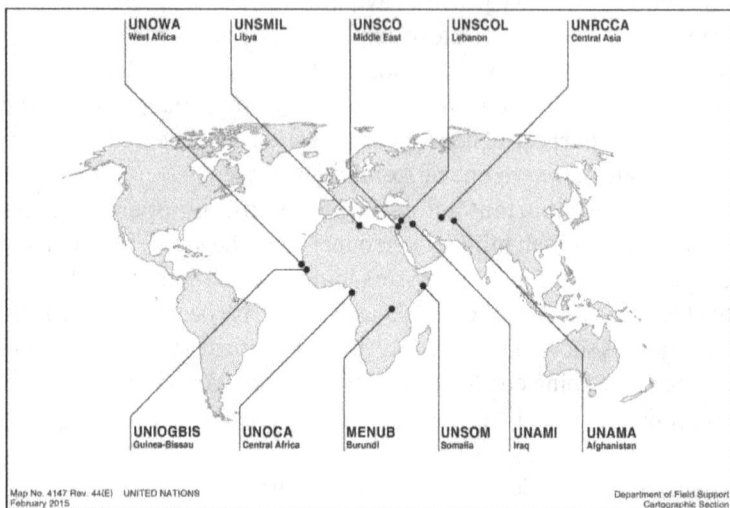

UNOWA West Africa	**UNSMIL** Libya	**UNSCO** Middle East	**UNSCOL** Lebanon	**UNRCCA** Central Asia

UNIOGBIS Guinea-Bissau	**UNOCA** Central Africa	**MENUB** Burundi	**UNSOM** Somalia	**UNAMI** Iraq	**UNAMA** Afghanistan

Map No. 4147 Rev. 44(E)   UNITED NATIONS
February 2015

Department of Field Support
Cartographic Section

NUMBER OF MISSIONS 11

PERSONNEL
Uniformed personnel 838
International civilian personnel (31 July 2015)..950
Local civilian personnel (31 July 2015)...1819
UN Volunteers 94
Total number of personnel serving in political and peacebuilding missions 3701

Figure 18.2: United Nations Political and Peacebuilding Missions
Source : http://www.un.org/Dpts/Cartographic/map/dpko/D_P_A.pdf (Factsheet: November 30, 2015)

3. 'Liberalism' (or democratic peace theory) is the school of
   international relations theory that aims to establish a 'stable
   peace', a state in which there is no risk of use of force, by
   changing countries from within into peaceful states through the
   propagation of the rule of law, democracy and a market economy.
Needless to say, I am fully aware that these theories on peacebuilding
measures are not perfect and in order to fix their flaws, we need to
adopt measures that integrate different approaches. The best example
where this has worked well is the peace zone established between
constitutional democratic states of Western Europe, North America,

Northeast Asia and Oceania. Nevertheless, even inside this zone where there is a common ground between states, i.e. constitutional democracy, it does not necessarily mean that they share the same national interests. Therefore, we need to keep in mind that regional disputes and conflicts are, to a certain extent, unavoidable.

Conclusion

The 'Introduction to the United Nations', published by the United Nations Department of Public Information, that I have relied on in writing this paper, explains that, based on the experience of peacekeeping/building operations, it is essential to invest all the necessary resources for supporting the development of the economy, social justice, human rights and proper governance in every country in order to achieve a sustainable peace.

Indeed, I have presented a pessimistic perspective on peacebuilding theories. However, before closing this chapter, I must confess that, given the various domestic and international disputes and conflicts around the world, even a pessimistic person like me shares the same idea emphasized in the above Introduction to the UN.

19 Green Growth/Green Economy: The New Economic Development Paradigm?

Dimiter Ialnazov

Introduction

The concept of green growth/green economy (hereafter, GG/GE) has been at the center of many international and national debates since 2008–2009. That was a time when three very important events took place. First, the global financial crisis was spreading around the world, threatening to exacerbate problems such as poverty and inequality. Second, in spite of the fact that humankind was not living within the planet's ecological limits, the world's leaders failed to agree on coordinated climate action at the 2009 UN Framework Convention on Climate Change (UNFCCC) negotiations. Third, in 2008 the EU adopted its energy and climate change package[1], taking the lead in reducing greenhouse gas (GHG) emissions and in promoting renewable sources of energy. In addition, the energy and climate change package was considered a vital part of the EU's Europe 2020 growth strategy.

The above events present a very short summary of the historical background to the revival of the interest of international organizations and national governments in the concept of GG/GE. As a result, we have witnessed a proliferation of reports by international organizations promoting GG/GE, such as, for example, UNEP (2011), OECD (2011b, 2013), the World Bank (2012) and UNESCAP (2012). Furthermore, in 2012 the Global Green Growth Institute (GGGI) became an international treaty-based organization that has been involved in the adoption of national green growth plans in sixteen emerging and developing countries.

The concept of GG/GE is generally thought to have three main applications: (1) to deal with global issues such as climate change, resource depletion and socioeconomic development (e.g. poverty and inequality); (2) to meet the challenges of sustainable development in advanced countries; and (3) to guide national growth strategies in emerging and developing countries. This chapter mainly addresses the third application, or the question of whether GG/GE can become the new paradigm of economic development. However, I should add that the adoption of GG/GE strategies by emerging and developing countries is also regarded as a necessary condition for dealing with climate change and resource depletion at the global level (that is, the first application).

The study of GG/GE is also very important for the goal of this book, i.e. to establish a new integrated academic field called Human Survivability Studies (HSS). As discussed in the Introduction of this book, HSS deals with complex social issues such as population increase, the spread of infectious diseases, environmental destruction, securing food, water and energy, correcting disparity and eradicating poverty. Since the international and national actors promoting GG/GE aim to tackle the problems of climate change, resource depletion, poverty and inequality, there should clearly be many things in common with HSS. Thus, the similarities and differences between research on GG/GE and HSS should be clarified.

This chapter is structured as follows. In the first part, I briefly review the conventional and modern views of development. I then analyze the concept of GG/GE in more detail and evaluate its benefits and shortcomings. Finally, I draw conclusions and delve into the issue of the similarities and differences between research on GG/GE and HSS.

Conventional and modern views of economic development

To put it simply, the conventional view is that economic growth, in particular the increase of GDP or GNI per capita, is the key to development. A developing country should expand its output at a rate faster than the growth rate of its population for a sustained period of time. Moreover, to catch up with advanced countries, a developing country should grow at a faster rate for a sustained period of time. Some famous examples from East Asia of successful leveling with advanced countries are Japan, South Korea and Taiwan in the post-war era, as well as China since the start of market reforms in 1979.

Furthermore, the following three characteristics should be added to the summary of the growth-based model of development: (1) the belief that economic growth automatically leads to poverty reduction and lower inequality (the so-called 'trickle-down effect'); (2) disregard for the adverse impacts of economic growth on the natural environment (increase of greenhouse gas (GHG) emissions and pollution, resource depletion and biodiversity loss); and (3) the idea that government intervention should be limited to facilitating the functioning of the market mechanisms and the provision of some basic public goods. To make a contrast between GG/GE and the growth-based model of development, I also call the latter 'brown growth/brown economy'.

Since the 1970s, there has been growing dissatisfaction with the conventional view of economic development. During the 1970s–1980s, most of the criticism pointed to the fact that in many developing countries the increase of GDP or GNI per capita did not actually lead to poverty reduction and lower inequality. The predominant way of thinking started to change in the direction of seeing development as something broader and much more complex than the quantitative expansion of average per capita income. The changing views of economic development led in the 1990s to innovations in terms of methods to measure it in a more holistic and integrated way. An example is the Human Development Index (HDI) that includes indicators of health and education[2].

During the 1990s, we have witnessed another wave of criticism aimed at the conventional view of economic development. Two main proponents of this wave were Amartya Sen and Partha Dasgupta. These scholars espoused a much more qualitative approach to development that understands it as improving quality of life, or human wellbeing. In its 2009 report, the Commission on the Measurement of Economic Performance and Social Progress (also known as the Stiglitz-Sen-Fitoussi Commission) gave a multidimensional definition of wellbeing based on the work of Sen, Dasgupta and other scholars. Apart from the material aspects such as income and consumption, the Stiglitz-Sen-Fitoussi Commission identified seven other dimensions: health, education, personal activities, political voice and governance, social connections, the quality of the environment and insecurity (Stiglitz, Sen and Fitoussi 2009: 14).

Another innovation was the inclusion of a concern not just about current but also future wellbeing. This was done by borrowing

Dasgupta's idea about wealth as the sum of three types of capital assets: the stocks of manufactured capital (for example, machinery and equipment), human capital (knowledge and skills) and natural capital (natural ecosystems)[3]. National wealth reflects a country's capacity to sustain both present and future wellbeing. A country's wealth could decline if that country destroyed/degraded its natural capital at a rate faster than that at which it accumulated manufactured and human capital (Dasgupta 2002: 3).

The inclusion of a concern about future wellbeing places sustainability at the heart of the new paradigm of economic development. According to the famous definition of the UN's World Commission on Environment and Development (also known as the Brundtland Commission), sustainable development is 'development that meets the needs of the present without compromising the ability of future generations to meet their own needs' (WCED 1987). The second part of the definition means guaranteeing that the stocks of resources passed on to future generations are not depleted by the current generation. Another way of putting it is that we should not consume resources today beyond the environment's carrying capacity.

Here a number of questions arise. For example, how can we estimate whether or not we are handing down a non-depleting stock of resources to future generations? Or, how can we measure sustainability? How can we determine the value of natural ecosystems when most are not traded on the market? Can the depletion of natural capital be compensated by increasing the sum of manufactured and human capital?

Since the early 2000s, a wide variety of sustainability indicators have been developed, and a comprehensive summary has been published elsewhere (see, for instance, Stiglitz, Sen and Fitoussi 2009: 61–82). The conclusion is against the formation of a single composite index and in favor of a hybrid approach that combines a) the monetary valuation of resources that can be traded on the market, and b) a set of physical indicators following the changes in the quantity and quality of various natural ecosystems that are difficult to measure in monetary terms. This appears to be an attempt to achieve consensus between 'weak' and 'strong' sustainability perspectives[4].

Finally, one should not forget the first part of the definition of sustainable development, which is about meeting the needs of the present generation. This signifies that we have to make a special coordinated effort today to improve human development around the world. An example of such effort at the international level is the UN agreement on the Millennium Development Goals (MDGs) in 2000. Although uneven and fragile, there has been substantial progress toward achieving some of the eight goals and twenty-one targets[5]. After the MDGs expired in 2015, the UN adopted their successor, the Sustainable Development Goals (SDGs).

To summarize, the modern view of development differs from the conventional view in the following three points. First, it shifts the emphasis from economic growth and measures of per capita income to human development and quality of life, or wellbeing. Second, it places sustainability at the heart of the thinking about development and embraces the view that we should sustain both present and future wellbeing. Third, it recognizes that poverty reduction, lower inequality and environmental protection will not happen automatically. Humankind needs targeted policy interventions at the local, national and international levels to achieve development goals.

GE/GG and emerging/developing countries

GG/GE's potential for emerging/developing countries

As discussed above, since 2008–2009 international organizations have been among the major players in the field of GG/GE. They can also claim credit for two of the most oft-quoted definitions. For instance, the OECD defined 'green growth' as 'fostering economic growth and development while ensuring that natural assets continue to provide the resources and environmental services on which our wellbeing relies' (OECD 2011b). According to the UNEP, 'green economy' signifies 'improved human wellbeing and social equity, while significantly reducing environmental risks and ecological scarcities' (UNEP 2011).

On a closer look, it can be said that the OECD's and UNEP's reports about GG/GE share the following characteristics: (1) a concern about the flaws of the prevailing brown growth-based model that has failed to solve problems such as poverty, inequality and environmental degradation; (2) a belief in the possibility of 'win-win' solutions, i.e. that we can improve people's lives while reducing GHG emissions, pollution and resource depletion; and (3) a reliance on technological and market-based solutions to decouple economic progress from resource consumption. In particular, the concept of GG/GE assumes governments play a strong role in terms of encouraging private sector investment in green technologies and products ('green innovation'), leading to the creation of green jobs.

In fact, the idea that the relationship between the economy and the environment can be a positive sum game is not something new. Many scholars have pointed out that the concept of GG/GE is based on the theory of ecological modernization developed in the early 1980s in Germany (Mol and Sonnenfeld 2000). According to that theory, the greening of the economy could be achieved by making better use of resources (i.e. higher resource efficiency) and incorporating environmental considerations throughout the whole product life cycle, starting from its design and ending in its recycling. In addition, the theory of ecological modernization assumes that the decoupling of economic progress from resource consumption is possible within the capitalist system (Lidskog and Elander 2012).

As mentioned in the Introduction of this book, the applications of the GG/GE concept should be different for high-income economies on the one hand, and emerging and developing countries on the other. In spite of their high level of income, the former continue to consume resources excessively and are still responsible for a large part of the world's GHG emissions. These economies have to change their current patterns of production and consumption to bring them within ecological limits and to reduce inequality. They also have a responsibility to assist developing countries' green transitions[6].

As for emerging and developing countries, we need to examine whether the concept of GG/GE can really be useful for them – can it become the new development paradigm as stated in the title of this chapter? There are at least a couple of reasons why people in these countries are more skeptical than those in high-income ones towards the inclusion of environmental considerations in policymaking. One reason is the view that taking care of the environment is a luxury that only wealthy countries can afford. The other is the idea that GG/GE is some kind of ploy by rich countries to hinder the development of poorer ones. For instance, there are considerable concerns that GG/GE might lead to 'green' trade barriers on exports from developing countries.

In comparison with the high-income economies, the emerging and developing countries have a relatively smaller ecological footprint, but face much greater challenges in terms of reducing poverty and inequality and ensuring people's access to health, education, electricity, safe water and sanitation. However, this does not mean that they can afford to disregard the adverse impacts of economic growth on the natural environment. Unlike the process of economic development in the nineteenth and twentieth centuries, emerging and developing countries today can no longer follow the model of 'grow (pollute) first, clean up later'[7].

The main reason for this is that due to population growth and the expansion of the middle class in emerging and developing countries over the next thirty to forty years, we can expect further large-scale increases in GHG emissions and resource consumption on a global scale. If we allow this 'business-as-usual' scenario to materialize, the increase of global average temperatures could reach four degrees Celsius above pre-industrial levels by the end of the twenty-first century (New et al. 2011)[8]. The results of such large-scale global warming could be catastrophic for the future of our planet and for the survival of humankind.

One may ask why emerging and developing countries should care about global issues such as climate change and resource depletion. Actually, many studies show that emerging and developing countries will be the ones most affected by these issues (see, for instance, UNFCCC 2007). One reason is their relatively higher dependence on agriculture that will likely be very seriously affected by the rise in global average temperatures. In addition, these countries also have a larger proportion of poor people, and there is evidence that shows that the poor are hit hardest in extreme weather-related disasters. The world has already seen a preview of what can happen if there is a big increase in food prices: during the 2007–2008 food crisis it was mostly emerging and developing countries such as Indonesia, Mexico and Egypt that experienced popular unrest and riots.

In a nutshell, emerging and developing countries cannot ignore the concept of GG/GE and regard it as something that should only be adhered to by wealthy countries. In its definition of 'inclusive green growth', the World Bank stated that developing countries' urgent need for rapid growth and poverty alleviation should be reconciled with the need to avoid irreversible and costly environmental damage (World Bank 2012: 2). Furthermore, the World Bank's experts warn that if developing countries do not embark on promoting GG today, their economies could be locked into unsustainable patterns, or they might risk facing costly policy reversals in the future.

Evaluating GG/GE: Lessons from emerging/developing countries

Since the practical application of the GG/GE concept has only recently begun, it is probably a bit early to give a definitive evaluation of its benefits and shortcomings. Therefore, I summarize the optimistic and pessimistic views on GG/GE and review the preliminary evidence of its success or failure based on several case studies.

Let's start with the benefits, as argued by the optimistic view. First, the scholars and experts in favor of GG/GE emphasize that, in spite of its shortcomings, our goal should be to eradicate the 'brown growth/brown economy' model that has failed to deliver progress in human development and environmental sustainability. There is a certain sense of urgency, or an awareness that we can no longer afford to postpone the solution of global issues such as poverty, inequality, climate change and resource depletion. GG/GE may not be the 'first best' way forward, but it promises to be at least a bit better than our 'brown' past.

A second benefit is that GG/GE serves to operationalize the concept of sustainable development. Another way of putting it is that GG/GE is a strategy towards achieving sustainable development in practice. The rationale behind this is that in spite of the broad consensus in favor of sustainability, little has been done to translate its principles into reality. As quoted in Lidskog and Elander (2012: 413), 'sustainability is so ambiguous that it allows actors from various backgrounds to proceed without agreeing on a single action'[9].

In contrast with sustainable development, we know more clearly what policy actions we must take to facilitate the green transition. A short list of GG/GE policy actions will certainly include increasing resource efficiency, shifting to renewable sources of energy, managing natural resources in a sustainable manner and investing in green technologies and products. Moreover, since powerful national and international actors have been pushing forward the GG/GE agenda, the expectations are that this time things will be different and we will see implementation of the ideas in practice.

Finally, OECD experts argue that GG/GE could be beneficial especially for emerging and developing countries because natural assets tend to be relatively more important for them in comparison with OECD economies (OECD 2013: 33). Natural assets play a significant role not only because exporting them elsewhere can bring a substantial amount of income, but also because they provide ecosystem services (e.g. clean air, drinkable water, CO_2 absorption, etc.) that are vital to sustain quality of life, or wellbeing. However, the management of natural assets such as forests, fisheries and freshwater is far from sustainable, implying the existence of potential threats to the wellbeing of poor communities.

One of the success stories among the GG/GE initiatives in the developing world is the Humbo Assisted Natural Regeneration Project in Ethiopia, which is directly related to the topic of natural asset (forest) management (World Bank 2012, citing Brown et al. 2011). As a result of deforestation and soil erosion, droughts and floods started to occur frequently and the productivity of agriculture declined. Since 2006, nearly 3,000 ha of forest has been regenerated, and a part of it has been designated as a 'protected area'. The project outcomes are 'win-win' as the logic of GG/GE would suggest. On the one hand, the incomes of the poor communities living in the area have increased, and on the other, the regenerated 'forest now acts as a carbon sink, absorbing and storing nearly 0.9 million tons of CO_2 over the project life' (World Bank 2012: 122).

Another success story is the Rural Electrification and Renewable Energy Development Project in Bangladesh. Like in other parts of the developing world, the problem is that many people in rural areas of Bangladesh lack access to electricity. To light their homes and cook after dark, people use kerosene fuel. Kerosene can be damaging for health and is also not cheap. The solution was found in installing off-grid solar home systems (SHS) to supply rural poor households with electricity generated from a renewable energy source (see, for instance, OECD 2013: 49–50). The project outcomes are again 'win-win' because of the numerous benefits for the rural poor and because of the positive effect on the environment in the form of lower pollution and GHG emissions.

Next I examine the shortcomings as argued by the pessimistic view. First, the fact that powerful national and international actors have been promoting the concept of GG/GE may be positive in terms of its chances of implementation, but there is also some criticism that decisions about GG/GE have been made in a top-down manner, without the participation of a variety of civil society actors.

According to its critics, GG/GE is just a new way for big corporations and financial institutions to continue making profits after the 2008–2009 crisis with the help of international organizations and national governments (Hoffman 2014). Therefore, we should not expect any significant progress in tackling the global issues mentioned above.

All this sounds quite similar to the way the concept of sustainable development has been rendered almost meaningless over the past twenty-two years[10]. As discussed earlier, in spite of the broad consensus in favor of sustainability, very little has actually been achieved. One possibility is that sustainable development has become 'a key strategy of sustaining what is known to be unsustainable' (Lidskog and Elander 2012: 413, citing other sources).

A second reason for the skepticism about GG/GE is its similarity with the ideas of ecological modernization theory. A great deal of criticism targets the 'economic growth' aspect of the GG/GE concept. As explained above, the modern view of development says that we should go 'beyond GDP/GNI growth' and focus more on quality of life, or wellbeing. In advanced countries, in particular, the idea that we should consume less and share more (through the 'sharing economy') has steadily been gaining popularity. For instance, the 'degrowth movement' argues that reducing consumption will actually increase our happiness and wellbeing[11].

According to another group of critics, due to GG/GE's reliance on technological and market-based solutions to decouple economic progress from resource consumption, the best scenario we can expect is one of incremental improvements rather than radical solutions (Lorek and Spangenberg 2014: 34; Hoffmann 2014). One argument in support of the pessimistic view is the existence of the 'rebound effect', where the increase of resource efficiency promised by the GG/GE proponents will actually lead to greater resource consumption and therefore further resource depletion[12]. Another argument is that GG/GE can only bring 'relative decoupling' (less GHG-intensive growth) but not a reduction of resource consumption in absolute terms (Lorek and Spangenberg 2014: 34; Hoffmann 2014).

Here the following question arises: if the GG/GE approach cannot offer solutions to global issues, then what is the alternative? What do the critics propose? Some (for instance, Lorek and Spangenberg 2014) argue in favor of a 'strong sustainable consumption' perspective, while others (Hoffmann 2014) point out that the transition to a sustainable economy would require a 'radical transformation of the capitalist system' as a whole. However, it is unclear what 'radical transformation' exactly means. Does it signify a new, post-capitalist system, or something called 'green or sustainable capitalism' (i.e. putting certain limits on capitalism's innate hunger for higher profits, while preserving its creativity and dynamism)? In other words, is 'green or sustainable capitalism' really feasible?

The final reason to be pessimistic about GG/GE is evidence from country case studies showing that its implementation in emerging and developing countries may be costly and therefore generate strong domestic resistance (Resnick et al. 2012). For instance, GG/GE may contradict traditional development strategies based on comparative advantage considerations, as shown in the cases of Mozambique and South Africa. Such strategies include deforestation to grow biofuels in the former and the use of coal for electricity generation in the latter. Although environmentally unsustainable, they have been supported by both the elites and the poor (Resnick et al. 2012: 216–218).

The evidence presented above remains incomplete and inconclusive. Yet, it may suggest that GG/GE could be successful at the level of individual projects (like those in Ethiopia and Bangladesh), but not at the national level. The reason is that the benefits of a national GG/GE strategy will be felt more in the long run, while in the meantime there will be short-term losers who may block its implementation. Therefore, GG/GE strategies in emerging and developing countries may not be feasible without strong international assistance to alleviate the short-term losses.

Conclusion

In this chapter I examined the question of whether GG/GE can become the new paradigm of economic development. Since 2008–2009, the concept of GG/GE has attracted a great deal of attention from international organizations and national governments, one reason being the possibility of its application to emerging and developing countries. The idea is that these countries can deal successfully with both socioeconomic problems (such as poverty and inequality) and environmental challenges at the same time. In addition, the adoption of GG/GE strategies by emerging and developing countries is regarded as a necessary step for dealing with the problems of climate change and resource depletion at the global level.

However, my conclusion is that we need much more empirical evidence to understand whether and how exactly the ideas of GG/GE could be useful for emerging and developing countries. On the one hand, OECD, UNEP and World Bank reports are full of success stories emphasizing 'win-win' outcomes, i.e. project results showing the simultaneous achievement of higher wellbeing and environmental benefits such as reducing GHG emissions, pollution and resource depletion.

On the other hand, there are also many critics who are skeptical about the benefits of GG/GE. The critics can be divided into two groups. The first group is composed of scholars and NGO activists who view GG/GE as just rhetoric aimed at helping big corporations and financial institutions from advanced countries to continue making profits in the wake of the 2008–2009 crisis. The second group includes scholars who try to evaluate the usefulness of the GG/GE approach based on empirical studies. According to one of these studies (Resnick et al. 2012), instead of being 'win-win', a GG/GE strategy at the national level is actually characterized by trade-offs between its socioeconomic and environmental goals. If implemented, it would lead to short-term costs, and among the losers we may find not just a few powerful cronies but also a large number of poor people.

What are the similarities and differences between research on GG/GE and HSS? First of all, we need to know more about HSS. As discussed in the Introduction of this book, HSS is a new integrated academic field that aims to develop the philosophy and methodology to solve complex global issues such as climate change, resource depletion, poverty and inequality. There is obviously an overlap in research themes, as GG/GE is also concerned with the same problems. In addition, as in the literature on GG/GE, HSS aims at developing a transdisciplinary perspective integrating insights from the humanities and natural and social sciences.

However, apart from the above similarities, HSS is quite different from the approach of GG/GE as it also includes research on the interconnectedness of the seemingly distinct global issues. An example is the interconnectedness of energy, water and food security, also known as the 'energy-water-food nexus'[13].

On a number of occasions the UN has pointed out that the lack of adequate access to energy, water and food to cover basic needs is the main barrier to overcoming poverty and inequality. In addition, we can predict that as a result of population growth, expansion of the middle class in emerging countries and climate change, the world's demand for energy, water and food will probably double by 2050, exacerbating the problem of resource depletion (see, for instance, REEEP 2014). Understanding and analyzing the vulnerabilities as well as the impending risks for individual countries, organizations and local communities cannot be carried out within the boundaries of traditional disciplines. We need a transdisciplinary, holistic approach that takes into account the above interconnectedness.

HSS aspires to offer such a holistic perspective based on the integration of individual disciplines from the humanities and natural and social sciences. To achieve this, HSS adopts an original methodology including forecasting, backcasting and meta-analysis. Furthermore, HSS relies on the synthesis of a wide variety of empirical case studies to identify patterns and formulate solutions.

In this chapter, I compared the ideas of GG/GE with the well-known concept of sustainability. I established that the thinking behind GG/GE has more in common with so-called 'weak sustainability'. On the other hand, HSS is more aligned with the stronger version of sustainability. Yet, as mentioned in the Introduction of this book, even though HSS and Sustainability Studies may seem similar, there are very important differences. The main difference is the emphasis in HSS on the sense of urgency, analyzing vulnerabilities and finding novel solutions to prevent or in response to crises.

Generally speaking, HSS has many things in common with the concepts of GG/GE and of sustainability. It does not contradict them, rather it complements them as it brings into focus some new perspectives and methods. Therefore, HSS can make an important contribution in support of our transition to a sustainable society.

20 Science and Trans-science

Eiichi Yamaguchi

Positioning of this chapter

The term 'trans-science' is a concept proposed by Alvin Weinberg (1915–2006) in 1972 and can be defined as 'Questions which can be asked of science and yet which cannot be answered by science' (Weinberg 1972). The word 'trans' is used as a prefix to represent questions that[1] 'transcend' science. The proponent, Weinberg, was an American nuclear physicist who was the Director of Oak Ridge National Laboratory from 1955 to 1973.

Originally, science is value-neutral, and accordingly, the truth discovered in science is the creation of new knowledge. Along with the creation of value by developing various technologies that utilize the knowledge built by the sciences, society has made the best use of it in politics and policymaking via the related decision-making process. However, there exists a field that is inseparable from both science and politics (decision-making of society), and this field is referred to as trans-science.

On March 11, 2011, there was a severe level 7 accident at the Fukushima Daiichi Nuclear Power Plant of Tokyo Electric Power Co. (TEPCO), caused by the tsunami that resulted from the Great East Japan Earthquake. In the wake of this event, the debate calling for a reconsideration of the relationship between science and society proliferated throughout Japan.

The discussion proceeded as follows. Since science did not originate in Japan, it would be better to entrust science to the experts known as scientists. However, science should then exist for the sake of society (science for society). In the case of bioscience, scientists should make 'human health' their objective instead of 'discovery' (turning discovery into health). Above all, if science can significantly damage society such as in the case of the accident at the TEPCO nuclear power plant, is it not necessary to resolve the problems democratically, which is the principle underpinning decision-making in modern society?

In other words, it is a paradox when a democratic society that advances based on majority rule, and science where majority rule does not hold any significance, occupy the same position. If we try to accommodate them forcibly, science would no longer remain science.

Therefore, this chapter summarizes Weinberg's initial argument and outlines its deployment in Japan. It also considers ways in which trans-science should be re-perceived so that a new relationship can be built between science and society. Moreover, by reconsidering innovation, which is another aspect of the relationship between science and society, we aim to come closer to a new understanding of trans-science.

Weinberg's argument

Weinberg classified the problems of trans-science into three types.

Type 1 includes those problems for which science cannot provide practical solutions, such as: 'the impact of low levels of radiation on living beings', 'the probability of occurrence of events that have extremely low probability' and 'engineering', etc. For example, it is impossible to determine the probability of simultaneous multiple emergency equipment failures in a nuclear reactor, and the fact that decisions should be taken by collating incomplete and fragmentary scientific information when implementing engineering in society, i.e. new technology, constantly exists. In any case, solutions can be deduced in the absence of financial and time constraints. However, it is difficult to find a solution in reality, and engineers have to make decisions by transcending science.

Type 2 includes those problems such as the 'conduct of humans and society', wherein the research subject entails a degree of uncertainty. Humans are not necessarily logical beings, and they perform a great variety of actions. Moreover, as a group, the actions of society are more uncertain, and it is impossible to predict the future even using the most advanced super computer. It is impossible for the social sciences to predict the actions of an individual or a society with the same level of accuracy as that found in the physical sciences. Therefore, social science (probably excluding economics) is indeed trans-science.

Type 3 includes the axiological problems with science. Originally, the priority of a field should be determined by evaluating the

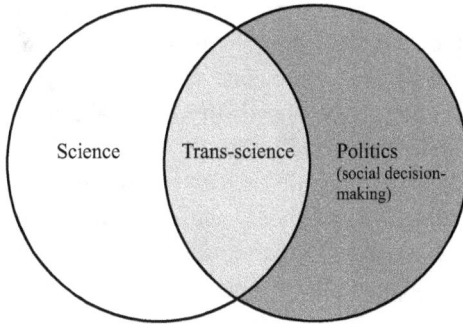

*Figure 20.1: Relationship between science and trans-science, based
on the work of Weinberg*

Source: Kobayashi (2007: 123).

'scientific value' of science that is meant to be value-neutral. Such 'ranking' to determine whether priority should be given to pure science or applied science, general science or specific science, taxonomy or paradigm disruption[2], search or codification, etc. is entirely based on ethics and intuition. In other words, since the 'value' of science needs to be tackled instead of the 'truth', decisions should be taken by surpassing science.

Further, Weinberg referred to the existence of 'The Republic of Trans-science'.

The validity of knowledge created by science is established and maintained in science, based on peer judgment and evaluation. On the other hand, there are no opportunities for the public to be involved in the evaluation of science created by scientists. This is exactly the 'Republic of Science' conceptualized by Michael Polanyi (1962). Scientists are the citizens of this republic, and accordingly, only scientists are allowed a voice. Credentials as a scientist are required for a person to be accepted as such in this republic.

On the contrary, 'The Republic of Trans-science' includes two kinds of members. One group comprises the citizens of the Republic of Science mentioned above, i.e., scientists. The other group contains all those involved in discussing the politics of this republic and making decisions (see Figure 20.1). The latter includes the public of a democratic society. However, participation is unevenly distributed away from the public in the case of undemocratic societies. Therefore,

Weinberg cites the following example to demonstrate that the existence of the latter is extremely important in terms of providing solutions to trans-science problems.

Discussions about the possibility of an accident at a nuclear power plant are frequently held in the US, and as a result, safety and emergency equipment are excessively installed to an extent that puzzles the experts. On the other hand, since the public of the Soviet Union did not have the right to participate in or be informed about such scientific and technical discussions, no containment vessel was installed in pressurized water reactors in that country.

Weinberg gave the following conclusion based on the above discussion. The roles and responsibilities of scientists in the discussions related to trans-science are to clearly articulate where science ends and where trans-science begins. Since scientists certainly know the location of the boundary of science, they have to instill their knowledge there in order to resolve trans-science discussions that tend to be confusing.

Further, scientists should welcome and encourage the participation of the public in this discussion. In the field of trans-science, scientific knowledge is not the exclusive possession of scientists. Since the trans-science republic encompasses both the science republic and civil society in which political decisions are made, its politics can be made to exist continuously in a democratic manner only through the latter, who can instill scientific knowledge into the trans-science republic to the greatest possible extent.

I have briefly explained Weinberg's argument above. Although it is a basic and elaborate discussion, I would like to point out its one flaw: he did not define the boundary between trans-science and science strictly. Although determining its position is the responsibility of scientists, since the method of distinguishing the boundary is entrusted to them, they are hesitant about forcing their way through the issues. Conversely, it leads to a situation where society does not encourage the participation of scientists, even if the problems of trans-science are manifest. I discuss this boundary issue further below.

Its deployment in Japan

Biologist, Atsuhiro Shibatani (1920–2011) was the first to respond to Weinberg's argument in Japan. He published the book 'Anti-science theory' in 1973, and after diligently introducing Weinberg's thesis, he supplemented the following:

> In the case of Japan... since the involvement of politics in the minute details of science is very easily approved, opportunities to carry out adequate discussions within science are limited. Consequently, the possibility of defining the boundary between science and trans-science in a straightforward way is also very limited, and it can be said that the opportunities for scientists to contribute to this from a neutral standpoint are rare. Instead, I ought to say that there were many cases in which even if a few scientists were to attempt such a thing, the remaining scientists would avoid or refrain from carrying out scientific discussions in public, or suppress them due to specific political influences or tacit solidarity. (Shibatani 1973: 168)

Whether willingly or not, scientists tend to have solidarity with political power in Japan, where politics easily becomes involved in the details of science. Since there is no opportunity to contribute from a neutral standpoint, the possibilities of defining the boundary between science and trans-science are limited. I argue that Shibatani's study predicted the fetters of a 'nuclear village' forty years ago, which was exposed by the accident at TEPCO's nuclear power plant in 2011.

Secondly, scientific philosopher Tadashi Kobayashi published 'The age of trans-science' in 2007 (Kobayashi 2007). In this book, along with a reevaluation of Weinberg's argument, he discussed the epochal significance of trans-science from the perspective of science communication, which is his specialty.

Kobayashi argued that traditional science communication can be described as the 'deficit model', which he attributes to the 'discrepancy between public opinion and that of specialists due to lack of scientific information'. Therefore, until now, science communication has been considered as a means to provide scientific knowledge to the ignorant public.

However, he advocates that instead of such one-way communication, the 'dialog model', in other words, the form of communication in which mutual learning can be achieved based on dialog between the public and specialists, is required. Additionally, he concludes by stating that, in the present where a trans-science-like situation is proliferating, the method of convening 'consensus meeting' between the public and specialists based on the 'dialog model' is certainly an extremely effective method.

Kobayashi continues by stating that there are conditions for accomplishing 'consensus meeting'. The first point is not setting consensus building as the goal. Since social antagonism might result from the antagonism between experts, informing the public that consensus cannot be established between specialists would only lead to confusion. Therefore, it is necessary to include all arguments and minority opinions. The second point is that the discussion must be 'rational'. Here, 'rational' indicates that the public would review the subject of discussion based on irrefutable common knowledge and their social experience. The third point is that the participation of experts from the social sciences or humanities is a must, as scientists from these backgrounds are required to separately process the discussion on an intellectual level in order to envisage the issues raised by the public.

Nonetheless, Kobayashi also argues that 'it is acceptable to give further consideration to the concept of civilian control of science and technology'.

> At present, although civilian control is an extremely important principle to control the rampage of military forces, on the contrary, it is also considered that it has a weakness in that the quality of military judgment would decline if the supreme commander were not an expert in military affairs. Even in this case, 'expertise' would be effective. Accordingly, the question is whether an ordinary person from a humanities background without any specialization in science and technology can regulate science and technology, and would the outcome be terrible?
>
> Still, I think that civilian control of science and technology is indispensable in modern society... It is necessary for persons from humanities backgrounds to obtain scientific literacy, and those from science and engineering backgrounds to obtain social literacy. Assuming this, two-way communication and civilian control is essential. (Kobayashi 2007: 83–85)

In Japan, the education system is differentiated into a humanities stream and a science stream from a considerably early stage in high school. Hence, students in the science stream loose the opportunity to study human psychology or social behavior in depth. In this way, scientists do not have the experience of studying the kind of impacts science has on society. Thus, even if a problem of the 'republic of trans-science' manifests, it is misread as a problem of the 'republic of science' and there is a risk of things getting out of control. For this reason, civilians are required to monitor and control the actions of scientists on the condition that those from the humanities stream acquire scientific literacy. Upon encountering the problem of trans-science, even though Weinberg persistently discussed scientists' code of conduct from the scientist's perspective, Kobayashi examined what civilians must do from the civilian perspective – this is where his originality on the subject of trans-science lies.

What is science?

This idea of 'civilian control of science' presented by Kobayashi has been accepted by many experts as a notable ideology since the accident at TEPCO's nuclear power plant in March 2011. However, I would like to object to this idea here. Although Kobayashi's argument seems to be 'fair' at a glance, upon returning to the true nature of science, this argument is capable of suppressing the evolution of science by diminishing the role of scientists.

Let's study 'what is science' in order to understand the reason and arrive at the truth.

In the book 'What is physics?', Sin-itiro Tomonaga (1906–1979) defined physics as follows: 'to pursue the laws of phenomena that occur in the nature surrounding us – however, mainly the laws pertaining to abiotic objects – based on observed facts' (Tomonaga 1979: 5). By naturally expanding this definition, physics can be defined as 'pursuing the laws of phenomena that occur in the nature surrounding us, based on observed facts'. Here, the issue is knowing what exactly it is 'to pursue the laws based on observed facts'.

In my book 'Five physics theories to learn before you die' (Yamaguchi 2014), I provide a detailed argument that the 'processes of Newton pursuing the law of universal gravity, Boltzmann pursing statistical mechanics, Planck pursuing the energy quantum hypothesis, Einstein pursuing the theory of relativity and de Broglie pursuing the concept of matter wave, have a common characteristic method of inference', while vicariously experiencing their mental processes. Furthermore, I express that this intellectual process, which is extremely essential for science, is not 'deduction' or 'induction', but the 'abduction'[3] that was first proposed by the American philosopher Charles Peirce (1839–1914).

The path of this process is described as:

1. Astonishing fact 'C' was observed.
2. If a certain hypothesis 'A' is correct, then 'C' would be its natural consequence.
3. Therefore, there exists a reason to consider hypothesis 'A' as correct.

The method of inference to deduce 'A' from 'C' is called 'abduction' (Pierce 1965). Furthermore, Pierce stated that 'all of the ideas of science come to it by way of abduction'.

On the other hand, 'deduction' is the method of inference where conclusion 'C' can be deduced inevitably if premise 'A' is temporarily deemed acceptable. Take the premise that a 'human is mortal'. If 'Socrates is a human', the conclusion that 'Socrates is mortal' can be inevitably deduced.

Further, 'induction' is the inference method of trying to discover universal law 'A' from an individual event 'C'. The opposite of 'deduction' is the 'generalization of embodied knowledge': 'Socrates, who is a human, is mortal'; 'Plato, who is a human, is mortal'. Therefore, the generalization is that 'All humans are mortal'. In this case, false conclusions are also obtained if there are exceptions.

If it is assumed that machines such as computers have reached a point where they can mimic human thinking based on big data, they would certainly be able to carry out 'deduction' and 'induction' more efficiently than humans. However, irrespective of the extent to which computers are developed, they can never reach the stage where they can carry out 'abduction'. This is undoubtedly the 'creation of knowledge', and only humans are capable of 'making what has never existed exist' and 'discovering what nobody knows'.

There is one more important concept for defining science – the confronting concept of 'night science' and 'day science'. Leo Esaki (1925–), who won the Nobel Prize in Physics for discovering the effect of quantum mechanical tunneling on semiconductors, said that science has 'two faces like Janus': a logos face and a pathos face (Esaki 1995). 'Day science' is ready and published in textbooks, etc., while 'night science' is nothing but tacit knowledge that is yet to be verbalized. A scientist repeats the process of searching blindly, trial and error, and occasionally discovers a breakthrough that shines light in the darkness. Esaki states that the bases of discoveries that move science forward have all been produced from such 'night science'.

Innovation diagram

Let us summarize the above discussion related to science.

There are two kinds of intellectual activity processes in humans: the 'creation of knowledge' and 'embodiment of knowledge'. The 'creation of knowledge' is not just 'discovering what no one knows or no one has ever seen'. It also signifies 'bringing into existence things that have never existed in this world'. This 'creation of knowledge' based on 'abduction' can be represented as 'research'.

On the other hand, 'embodiment of knowledge' is the intellectual activity of joining and integrating such newly created 'knowledge' mutually or with existing technology and bringing it to a level where it has economic and social 'value'. This 'embodiment of knowledge' based on 'deduction' can be represented as 'development'.

Here, let us try to draw a diagram as shown in Figure 20.2. Let's take 'creation of knowledge' as the x-axis and 'embodiment of knowledge' as the y-axis. Thereupon, the three intellectual activities – starting from existing knowledge and embodying that knowledge is 'deduction' (S→A), causing the opposite effect is 'induction' (A→S) and creating entirely new knowledge is 'abduction' (S→P) – can be represented in a single two-dimensional plane. As first proposed in my book 'Innovation: Paradigm disruptions and fields of resonance' (Yamaguchi 2005), let us call this diagram the 'innovation diagram'.

In this innovation diagram, let us draw a horizontal boundary between 'technology' to which economic and social values have been assigned, and 'knowledge' to which such values have not been assigned. The area above the boundary line is the former, and that below is the latter. The implications of this diagram can be experienced if the area below the boundary line is considered 'soil', and if the dynamics of a bud germinating within the soil is treated as the analogy of innovation.

'Abduction' is carried out entirely under the soil. This indicates that the process of knowledge creation is 'night science' and is undertaken in total darkness – to progress in total darkness without any light, manual or textbook. Here one has to move forward by relying solely on tacit knowledge. In this regard, the public, who

Knowledge
embodiment

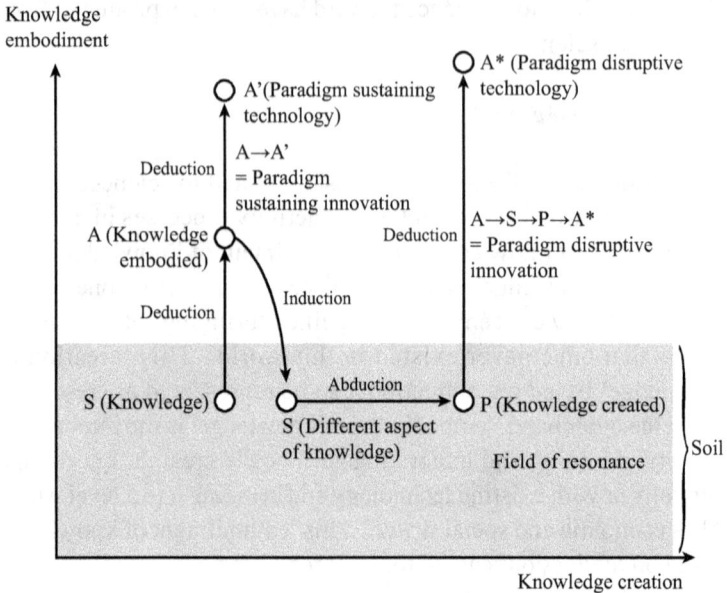

Figure 20.2: The innovation diagram

Note: Science is depicted under the soil (gray area), whereas trans-science belongs to the area over the soil (white area). The border between science and trans-science can clearly be defined as turning to deduction from abduction. Thus, the field of resonance gives rise to an essential role for accomplishing 'the Republic of Trans-science'.

carry out economic and social activities, live above the soil and cannot see 'under the soil' or, in other words, the 'night science'.

On the other hand, the sun shines above the soil, where it is possible to have an expansive view, and this is a world that can be seen from the perspective of the market. 'Day science' belongs to this area 'above the soil'. Here, value is created when a sprout germinates above the soil based on 'deduction'. In other words, tangible value is brought to the world in the form of a new product or service. That is to say, the 'embodiment of knowledge' can also be called the 'creation of value'.

Why is Figure 20.2 called the innovation diagram? Innovations including all kinds of reform activities that bring about economic and social value are produced from the chain of activities involved in the 'creation of knowledge' and 'embodiment of knowledge' ('creation of value'); it is possible to include all these chains in this innovation diagram. Originally, this diagram can be treated as an analogy for the flourishing of plants. As soon as a shoot protrudes out of the soil, it grows steadily towards the sky. However, this growth stops at some point; eventually, the plant's lifespan comes to an end and it withers.

Similarly, the process of 'deduction', in which the embodiment of knowledge is continued by aiming to improve the added value, invariably reaches a dead-end once it has undergone several repetitions. For instance, in the Silicon LSI (Large Scale Integrated circuit), which has been achieving an integration scale of four times every three years in accordance with Moore's law, the distance between electrodes towards the end of the year 2020 will be below ten nanometers. It follows that if further miniaturization is carried out, the electrons would start moving as a 'wave' with wave functions instead of particles, and the speed would no longer increase. Therefore, since there is no point in conducting further miniaturization, the existing semiconductor industry would reach its limit. Let us call innovation produced in this way, by repetition of 'deduction' on a paradigm 'S' that eventually reaches a dead-end, 'paradigm sustaining innovation' (A→A').

What happens once a dead-end is reached? Trees would shed their seeds onto the soil to create new life, and bamboo would grow rhizomes under the soil to create new bamboo shoots. If the soil were rich, the flourishing root or rhizome would plow through it and give birth to a new sprout above the soil. Similarly, the process temporarily distills to its true essence by carrying out 'induction' (A→S). Thereupon, 'abduction', i.e., 'night science', becomes possible under the soil, and a new paradigm can eventually be attained (S→P). In this way, with the 'embodiment of knowledge' or the growth of the sprout of 'deduction', it is possible to arrive at an entirely new value (P→A*). Let us call such an innovation of 'induction'→ 'abduction'→ 'deduction' through soil as 'paradigm disruptive innovation' (A→S→P→A*). Moreover, let us call the node 'P' a 'field of resonance'. This is the field in which resonance is produced between the desires of those seeking existential desire in the 'creation of knowledge' with those seeking existential desire in the 'embodiment of knowledge', even while they each recognize the differences in the other's desires.

By extending the two-dimensional innovation diagram of 'creation of knowledge' and 'embodiment of knowledge' and making it three-dimensional with the addition of another dimension known as 'Crossing the frontiers of knowledge', as shown in Figure 20.3,

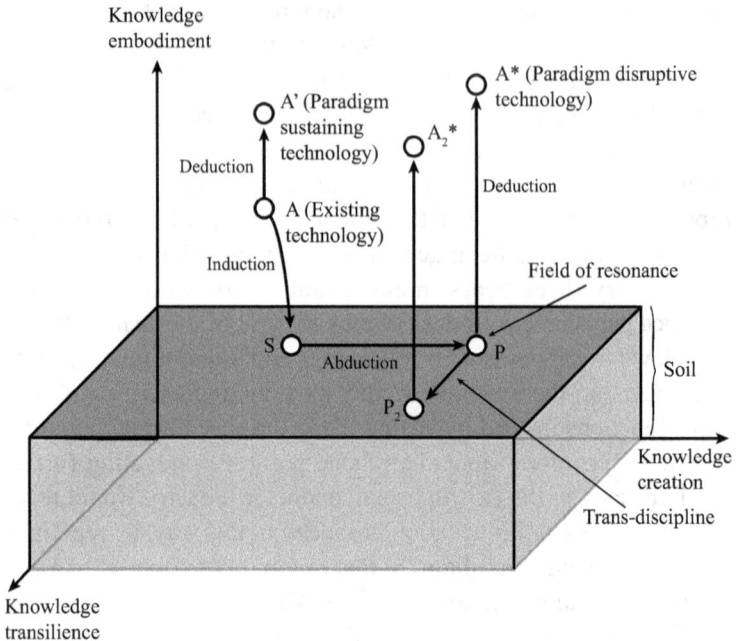

Figure 20.3: Three dimensional representation of the innovation diagram

Note: The three dimensional space is spanned by knowledge creation, knowledge embodiment and knowledge transilience. In the diagram, vectors describe four methods of human inference: deduction, induction, abduction and trans-discipline.

it is possible to deepen the innovation diagram's significance. For example, similar to the birth of the central dogma of molecular biology with the discovery of the double-helical structure of DNA by James Watson (1928–) and Francis Crick (1916–2004), who crossed the frontiers of biology from physics, extreme paradigm disruptions are occasionally created by crossing disciplinary boundaries. Furthermore, scientists can obtain entirely new clues or ideas from the knowledge acquired by philosophers over time. In this way, let us term this intellectual activity of practically accomplishing 'crossing the frontiers of knowledge' from physics to biology and further to philosophy as 'transilience'. For humans, this fourth type of inference method would mean 'leaping into a completely different evaluative space'.

Recapturing trans-science

Let us try to read once again 'The age of trans-science' based on the understanding depicted in this innovation diagram. When we read it, we notice that Kobayashi does not discuss 'what is science' even once, and in addition, he proceeds with the discussion by confusing science (creation of knowledge) and technology (embodiment of knowledge and creation of value), and develops a chunk of 'science and technology'. We do not even give a passing thought to the fact that scientists pursue breakthroughs in absolute darkness, observed and seen by nobody, and do their best to bring into existence those things considered impossible. In other words, true science, i.e. 'night science' is entirely absent from the 'science and technology' he mentions.

So, starting from the innovation diagram described in the previous section, I would like to properly define the boundaries of science and trans-science that went undefined in the Weinberg debate. From that perspective, reconsidering the association between science and society, the 'consensus meeting' with the public and scientists should look completely different from the conflicting perspective of 'civilian control imposed on science and technology'.

Let us make several assumptions, as follows. In Figure 20.2, the soil surface itself represents the boundary between science and trans-science. It is assumed that 'night science' under the soil is science, and 'day science' above the soil is trans-science. Essentially, the world of 'creation of knowledge' exists under the soil without the existence of 'value'. This is consistent with the principle that the emergent 'knowledge', which is value-neutral from science and pays no regard to 'truth, goodness and beauty', proves to be an asset common to humankind worldwide. On the other hand, the world of 'value' above the soil is created by 'deductively' realizing the 'knowledge' born in this manner.

By defining the boundaries of science and trans-science in this manner, we can dispel a variety of misunderstandings in the following manner and can also avoid unnecessary confusion.

Firstly, we can prevent the 'withering of scientists and subsequent suppression of science' as discussed above.

As previously mentioned, in the world of 'night science' under the soil, since the emergence or seeds of a variety of hypotheses and discoveries have not yet been verbalized while retaining their implicit knowledge, scientists have to trust their own senses and imaginations and proceed while searching blindly through the darkness. Once the public understands that such a world of 'abduction' exists, they will be able to perceive how scientists work and understand the significance of fostering social investment in scientific pursuits. Since the presence of 'night science' itself brings about paradigm destroying innovation, society will stop dismissing it as 'useless', driven as it is by the curiosity of scientists.

Secondly, scientists will be able to guarantee impartial contribution to the challenges of trans-science. As argued by Shibatani (1973: 164–169), in Japan, since the boundary between science and trans-science was not properly defined, politics could easily transgress the inner workings of science. Therefore, inexorably, scientists often had to support political motivations. However, if society recognizes the social significance of 'night science' and also, if the non-intervention of politics into this aspect of science is guaranteed, scientists will be able to discuss the challenges of trans-science even more aggressively.

Thirdly, it is possible to shift the conflicting relationship between society and science towards one of harmony.

Weinberg argued:

> The republic of trans-science, bordering as it does on both the political republic and the republic of science, can be neither as pure as the latter nor as undisciplined as the former. The most science can do is to inject some intellectual discipline into the republic of trans-science; politics in an open society will surely keep it democratic. (1972: 222)

As he claims, if a trans-science problem manifests, scientists are obliged to reach out proactively and speak out without bias or favor. The existence of 'night science' is the only thing that guarantees this impartiality.

I go one step further and imagine the arrival of a society of citizens who actively visit the world of 'night science' by becoming 'people of science' to the maximum extent possible. This relates to research that exceeds the arts and sciences and involves intelligent activities carried out across various stages of life that include seeing things that nobody has seen and bringing non-existing things into being. This means striding into 'night science' in order to stretch and elevate one's 'innermost feelings'. In this way, professional scientists and citizen scientists will understand the purpose of each other's lives and can build a 'resonant world', which will certainly bring a new dimension to the relationship between science and society.

Two case studies

Based on the above discussion, let us reconsider two trans-science cases that have occurred in Japan in recent years. First is the accident that occurred on JR Fukuchiyama line on April 25, 2005 at 9:18am.

A rapid southbound train that started from Itami station sped on a straight line railway track for 6.5 km and derailed on a right curve between Tsukaguchi and Amagasaki stations. In this crash, 107 people died and 562 people were injured. As has been evidenced in my book 'The essence of the JR Fukuchiyama train incident' (Yamaguchi and Miyazaki 2007), when the design of the curved route was changed from 600 m radius to 304 m radius in December 1996, JR West Japan did not install an ATS (Automatic Train Stop) and did not seek an overturn speed limit. Thus, the occurrence of overturning was almost predetermined with a probability of one. This was the cause of the crash. It should be noted that physics proved that the probability of overturning could have been zero if the 600 m radius had been retained.

However, the Accident Investigation Commission (under Chairman Goto Norihiro) published a final report in June 2007, which mentioned that this accident occurred because 'the train driver delayed the application of brakes'. After having concluded that the train driver was fully responsible, random guesswork was carried out 'regarding the delayed application of the brakes by the train driver'. It was surmised that 'the train driver thought that the in-train phone call seeking a false report was cut, and the distracted train driver who was attentively listening to the telecommunication between the conductor and transport commander lost his focus on driving'.

After that, a number of criminal trials were held. In the first criminal trial against Masao Yamazaki, who was the General Manager of the railway during the time when route design changes were determined, the Kobe District Court issued a not guilty verdict, as provided below, in January 2012.

The train driver drove the train over the curve at a speed that exceeded the overturn speed limit, which lead to the overturning and derailment of the train causing passenger casualties. It would be difficult if one denied that such things could occur within the range of predictability just by saying that 'this happened due to some reason' or 'these things are destined to happen'.

However, since the circumstances leading to the exceedance of the train overturn speed limit, which is considered to be a target of predictability, are vague, the possibility of the occurrence of results is also not specific. In this context, if predictability means the occurrence of results within the predictable range, then in essence, this is not very different from the sense of fear. Therefore, there is no option but to say that the prediction of result occurrence is not easy and the extent of predictability is considerably low.[4]

Readers will easily see that in this judicial decision the judge dismissed the science out of hand. Science sought the accurate overturn speed limit. Thus, it was easily possible to quantitatively predict. Moreover, the actual overturn speed concorded with the overturn speed limit that was determined theoretically (106 km/h). In December 1996, when the radius of the curve was changed from 600 m to 304 m, the occurrence of overturn accident was finalized with a probability of one, thereby ensuring an enormously higher degree of predictability. The science used here was high school level physics, which all the railway managers should have been familiar with.

Why did the train driver end up in an undecidable situation on the straight line track (speed limit 120 km/h) that continued for 6.5 km in the lead up to this curve with a radius of 304 m? Why did the managers block the installation[5] of an ATS on the line? Why didn't the managers know the overturn speed limit? Nobody can understand why the court did not attempt to answer these three questions.

This first example can be considered a typical case showing that though this is essentially a trans-science issue, the scientists and judicial authorities failed to share their knowledge and carry out reasonable interactions and discussions with each other. If this judgment is absorbed as a legal precedent in Japan, a bizarre country will be born where 'persons, who are not scientifically literate and who do not even know the meaning of physical limits, will be appointed as the representative of organizations, and eventually, all kinds of organizational accidents will be exempted'.

The second example is the TEPCO nuclear accident that occurred between March 11 and 14, 2011.

As of November 2014, the then three executives, namely, Chairman Katsumata Tsunehisa, Vice President Muto Sakae and Takekuro Ichiro (Fellow) were prosecuted by the Committee for the Inquest of Prosecution for 'not taking appropriate action even when a report had been received on the possibility of a tsunami of height fifteen meters or more'.

Whether appropriate tsunami mitigation actions were taken or not should be considered under 'risk management', and this resembles the situation in the Soviet Union mentioned above as described by Weinberg. The problem as to why Japan's risk management is far out of line with its status as a democracy is worthy of consideration. Since many studies have already discussed this problem, let us address another issue in this chapter – technology management. The problem here is whether or not the executives put in their best managerial effort to prevent the three power stations from falling into crisis after the occurrence of the tsunami.

Regarding this, as has been confirmed for the first time in my book 'FUKUSHIMA report: Nature of the nuclear accident' (Yamaguchi, Nishimura and Kawaguchi 2012), Takekuro Fellow rather intentionally obstructed the best possible technological management strategy and allowed the reactor to run beyond its physical limits. A technological management strategy that does not allow for the exceeding of physical limits does nothing but keep the reactor water level on the plus side. The then Prime Minister Kan Naoto and Prof. Yasushi Hibino suggested that while the reactor water level is somehow maintained in the plus level through RCIC operation (cooling system used for isolation), opening the vent and the immediate injection of seawater must be carried out. However, Takekuro, who was in the Prime Minister's Office as a management representative of TEPCO, continued to deliberately refuse this process, as he did not want to perform decommissioning. Runaway could have been avoided if TEPCO management had been science-literate to at least know that as soon as the reactor crosses a physical limit that is based on the laws of physics, it goes out of control beyond human understanding.

Here, the problem is that in spite of graduating from a Faculty of Engineering (Department of Mechanical Engineering), why did Mr. Takekuro refuse to carry out seawater injection? After joining TEPCO, he traversed the field of nuclear power consistently and served as Kashiwazaki-Kariwa Nuclear Power Plant Superintendent,

Engineering Research and Development Deputy Chief Nuclear Officer (CNO) and as the Vice President and the General Manager of the Nuclear Power and Plant Siting Division, which is the apex of the nuclear field. Therefore, he clearly must have been well versed in the detailed technology of the pipe structure of the nuclear reactor. However, he lacked knowledge in physics and was ignorant regarding the physical limit, which forms the basis of nuclear reactors.

In other words, the problem presented in this second example does not stop at the divide between arts and sciences. Rather, it represents a typical example, wherein the refusal to transgress the boundaries of knowledge between different fields of science complicates the resolution of trans-science issues.

Conclusion: Become citizen scientists

For the purpose of resolving the issues of trans-science, I sincerely hope that a citizen scientist society is realized where knowledge is liberated and every citizen becomes a person of science. Everyone will agree that science progresses rapidly and eventually splits and specializes into different fields. It is, therefore, no longer possible to learn and grasp the whole picture. However, I firmly believe that this is wrong. In science, the basis of its life form can be understood by anyone by learning the process of abduction of those who have created it. The basis can be comprehended, even when the variety of techniques and terminology derived from it are not understood. In that sense, Kobayashi's message that 'scientific literacy is required in the arts and social literacy is required in the sciences and engineering' and 'bi-directional communication is necessary upon obtaining that' is correct. We should harmonize science and society and create a new knowledge to resolve trans-science issues as quickly as possible while transgressing the boundaries of knowledge.

The question is, how to create that transdisciplinary society?

Firstly, based on the vision outlined here, a graduate school needs to be established where the bases of arts (humanities and philosophy, economics and management, law and politics, cross-cultural communication) and science (science and technology, pharmaceutical and life sciences, information and environment), are broadly studied. At such a graduate school, scientists, social scientists and humanities scholars should work together as lecturers in the same 'fields of resonance'. It is needless to say that the students who study there will earn their PhDs and claim their own 'abduction'.

Secondly, whenever a social issue appears to be associated with human survival, as indicated in the three-dimensional innovation diagram (Figure 20.3), we should ensure 'transilience' while freely transgressing the boundaries of knowledge from physics to philosophy and resolve the issues by resonantly drawing on wisdom from a variety of fields. The challenges that need to be dealt with using the new knowledge are as follows: to establish how a Fukushima earthquake will recur, to scientifically plan long-term energy policies and to obtain a national consensus on that scenario.

The Japanese people, who posed a serious threat to survival in the world by causing a nuclear power accident, have a responsibility to build that kind of graduate school. A country develops into a bizarre nation if persons, who are not science literate, are appointed to responsible positions, and all kinds of organizational accidents are exempted owing to that. Japan should never become such a nation.

Epilogue: Our Responsibility to Future Generations

Masakatsu Fujita

'Life', or 'survival', is the topic addressed by Human Survivability Studies. This field of inquiry cannot, however, solely focus on 'life' and 'survival'. Only in relation to 'death' and 'extinction' can the nature of life be elucidated and its significance discussed.

One characteristic found throughout our modern ways of living is that our attention has been captivated by life's splendor and we have stopped turning our gaze toward death, or that which imposes a limit on our lives. We are losing the humility that comes from looking directly at life's finitude. We pursue a comfortable lifestyle dictated by our desires, squandering resources and destroying the environment around us. These circumstances are inseparable from the fact that 'survivability' has now become a particularly salient issue.

As I discussed in Chapter One, with the remarkable development of science and technology we now possess the power to irreversibly destroy the environment, and our actions have come to affect not only the present but also the distant future. 'Survivability' is no longer an issue that concerns only those living today; it has become a problem that is profoundly connected to both the continued existence of the natural environment that surrounds us and the lives of future generations.

In the words of Hans Jonas (1979), we must pay attention not only to others who stand before us now but also to the Earth's environment as a whole and the rights and survival of future generations.

For example, in the modern era economic activities have been carried out on an unprecedented scale, consuming large amounts of resources and producing vast quantities of goods. It is unthinkable, however, that we should use all of the Earth's resources for our own generation, thus driving future generations into poverty. This would be an injustice against the people of the future.

Nuclear power generation, for example, creates radioactive waste that has the potential to cause great harm to human beings and the environment, and it is physically impossible to render this waste harmless or shorten its lifespan. Tens of thousands of years are needed for it to become harmless via natural processes. This radioactive waste is being provisionally stored, while the difficult questions of 'how' and 'where' it is to be dealt with in the long-term remain unanswered. Normally, it would be inconceivable to produce something without giving any thought to the waste it creates, but this is what is being done in the case of nuclear power. As a result, we are forcing the concomitant risks and economic burdens on future generations without their consent.

What must now be given particular consideration in relation to the question of 'survivability' is our responsibility to nature and to future generations. Responsibility is something that must be addressed not only in regard to our past actions and their consequences; the actions we are in the process of choosing right now, and the consequences they are likely to bring about, can and must also be taken into consideration.

As long as there is a chance that our actions will affect the Earth's environment as a whole, or the possibility that what we do today will have a critical impact on the lives of future generations, we have responsibilities and obligations we must acknowledge, even if we do not know those involved directly and cannot determine the consequences of our actions with absolute clarity.

In recent years there has been talk of 'sustainable development' that does not violate inter-generational parity. In reality, however, this idea has not been adopted with sufficient gravity. Fairness seems to have been donned as a disguise to facilitate the continuation of development. Right now, it is fair to say that preventing this idea from becoming simply an ad-hoc excuse is of the utmost importance. This is another reason why we must strengthen our awareness of our responsibility to future generations, choose our own actions accordingly and build a society of the future. These imperatives are profoundly connected to the direction 'in which we are headed', one of the core issues being examined in Human Survivability Studies.

Notes

Chapter 2

1 The radial velocity method is one of the principal techniques used in the search for exoplanets. It is also known as Doppler spectroscopy. Just as a star causes a planet to move in an orbit around it, so a planet causes its host star to move in a small counter-orbit resulting in a tiny additional, regularly-varying component to the star's motion. Jupiter, for example, causes an additional movement of the Sun with an amplitude of thirteen m/s and a period of twelve years. The Earth's effect is much smaller, amounting to fluctuations with a ten cm/s amplitude over a period of one year. If a star is accompanied by a planet, the radial velocity of the star will periodically change as the star moves toward and then away from a distant observer. If the effect of the planet is sufficiently large, through a combination of its mass and orbital radius, the oscillating movement of the star is detectable as a small periodic blue shift and red shift in the star's spectral lines. Historically, radial velocity measurements had errors of 1,000 m/s or more, making them useless for the detection of orbiting planets. However, beginning in 1980, Bruce Campbell and Gordon Walker developed a method capable of measuring radial velocities to a precision of fifteen m/s. A number of groups around the world are now using the radial velocity method, with a precision in the range of three to ten m/s in the search for extrasolar planets.

2 An Astronomical Unit (AU) is a unit of measurement equal to 149.6 million km, the mean distance from the center of the Earth to the center of the Sun.

3 'Solar system formation model' refers to the Kyoto Model. Large gas planets that are close to their star (~ five AU) are thought to be formed when gas accumulates around the solid core. The Kyoto Model was a solar formation model proposed by Professor Hayashi and his group at Kyoto University around 1970. Also called the smallest mass model, the hypothesis is that planets in the solar system are formed with the smallest mass. One problem is that when using the Kyoto Model for planets in the area of Neptune, the core formation is much longer than the age of the solar system; this is an issue with the core accretion model for Neptune (Hayashi, Nakazawa and Nakagawa 1985: 1100).

4 The Kepler space telescope examined around 145,000 main-sequence stars, and including all stars examined by 2016 the figure would amount to 200,000 in total. Our galaxy contains more than 100 billion stars.

5 In astronomy and astrobiology, the circumstellar habitable zone, or simply the habitable zone, is the range of orbits around a star within which a

planetary surface can support liquid water given sufficient atmospheric pressure. The name Goldilocks comes from the children's fairytale 'Goldilocks and the Three Bears', in which a little girl chooses from sets of three items, ignoring the ones that are too extreme (large or small, hot or cold, etc.), and settling on the one in the middle, which is 'just right' (see http://iopscience.iop.org/article/10.1088/2041-8205/767/1/L8/ meta;jsessionid=F69613E8B32AA5B5EE38364ECBAC6DFC. c2.iopscience.cld.iop.org).

6 In this paper, Professor Kazunari Shibata's group at Kwasan Observatory, Kyoto University have proven for the first time that solar-type stars have frequent and high energetic flare activities even without the accompaniment of large and short-frequent exoplanets (hot Jupiters).

7 A moist atmosphere (saturated by water vapor) has an upper limit of the outgoing infrared flux that can be emitted. If insolation exceeds this limit, temperature increases until complete evaporation of liquid water from the surface has occurred. The limit of the planetary radiation (longwave radiation) of a planet with oceans on its surface is determined by various mechanisms called 'radiation limits', which can be classified as the Komabayashi-Ingersoll limit and the radiation limit of the troposphere.

   The asymptotic value of OLR, 400 W/m^2, is extremely close to the Komabayashi-Ingersoll limit, which is determined from the constraint on radiation flux passing through the stratosphere in a 1D model. However, the value of the Komabayashi-Ingersoll limit, 385 W/m^2, is the limit for cases in which the tropopause is saturated. The tropopause was not saturated in the results of our 3D calculations and relative humidity at the tropopause for S1570 was approximately 0.5. At a value of relative humidity of 0.5, the upper limit of outgoing radiation that is determined by the constraint on radiation flux passing through the stratosphere will be approximately 450 W/m^2. (For more information, see 'The Komabayashi-Ingersoll Limit When Relative Humidity is Considered'.) This implies that the asymptotic value of OLR in the 3D model is not determined by the stratospheric flux constraint. If 3D results correspond to the 1D equilibrium solution, the asymptotic value of OLR and the constraint on the radiation flux emitted from the troposphere should be the determining factors. The feasibility of this hypothesis is examined in the next chapter (see http://www2.nagare.or.jp/mm/98/ishiwata/english/genkai1.htm).

8 The faint young sun paradox describes the apparent contradiction between observations of liquid water/liquid ocean early in Earth's history and the astrophysical expectation by standard solar model that the solar luminosity would have only been 70–75% as intense during that epoch as it is during the modern epoch. With such weak solar insolation, the Earth would become 'Snowball Earth' unless it had a much higher greenhouse effect than that of current Earth.

   In Airapetian et al.'s article, it was suggested that on an Earth with a faint young sun, the main greenhouse gas that sustains warm temperatures with liquid ocean on the surface of the Earth with weaker solar radiation was nitrous oxide, generated by CME (superflare) from the active young sun, not CO_2, CH_4, and N_2.

Chapter 5

1 The first edition of the *Handbook of Mixed Methods in Social and Behavioral Research* appeared in 2002 (Abbas Tashakkori and Charles Teddlie, eds). The *Journal of Mixed Methods Research* was launched in 2007 and the Mixed Methods International Research Association was inaugurated in 2013.

2 Although there is a difference in the meaning of 'transition to a market economy' and 'post-socialist transformation', in this chapter we use these terms interchangeably.

Chapter 7

1 'Informatics no. 1', Graduate School of Informatics, Kyoto University.
2 See http://en.wikipedia.org/wiki/Moore%27s_law.
3 See http://en.wikipedia.org/wiki/Global_surveillance_disclosures _%282013%E2%80%93present%29.
4 See http://www.dmat.jp/.
5 See http://www.jmooc.jp/.
6 See http://gacco.org.
7 From the opening remarks made at the First Congress of the Association of Space Explorers, held in Cernay, France on October 2, 1985.
8 See Wikipedia http://en.wikipedia.org/wiki/Search_for_extraterrestrial_ intelligence.
9 For more detail, see http://setiathome.ssl.berkeley.edu/.
10 See http://www.healthmap.org/site/about.
11 See http://www.sas.com/ja_jp/news/press-releases/2014/september/2014-09-03-shionogi-imstat-for-hadoop-jp.html, in Japanese.
12 See http://business.nikkeibp.co.jp/article/Big Data/20140422/263347/?P=3, in Japanese (no longer available online).
13 See http://www.asia.si.edu/collections/edan/default.cfm.
14 See http://www.soumu.go.jp/johotsusintokei/whitepaper/ja/h24/html/ nc254510.html, in Japanese.

Chapter 8

1 While shutdown due to global warming would not cause an ice age, as was depicted in recent blockbuster *The Day After Tomorrow*, eastern North America and Western Europe would nevertheless experience a climatic shift.

2 The River Law of October 7, 1964 (#167), last modified November 22, 2013 (#76).

3 Flood event of a given annual flood probability, which is generally taken as: fluvial (river) flooding likely to occur with a 1% annual probability (a one in 100 chance each year), tidal flooding with a 0.5% annual probability (a one in 200 chance each year) or estimated flood using a hydrological model assuming that the projective rainfall (with a Return Period of

100–200 years) occurred in the catchment without considering flood
controlling structures.
4 The Automated Meteorological Data Acquisition System operated by the
   Japan Meteorological Agency.

Chapter 9

1 Ethnic groups are communities distinguished from other communities by
   blood relationships, social structure and other common features.
2 Culture is the entire complex of abilities and customs shared by members of
   a particular society.
3 Religion is the belief in a supreme being that shapes one's understanding of
   life.
4 See '60 Ways the United Nations makes a Difference', published by the
   United Nations Information Centre.
5 Note that if we only look at family law cases, the total number of cases
   was 350,542 in 1989, meaning that it has increased 2.6 times in less than a
   quarter of a century.
6 See 'Overview of the Status of Human Rights Infringement Cases in 2015',
   published by the Ministry of Justice.
7 Namely, the Human Rights Bill of 2002 submitted by the government, the
   Bill of the Act on Relief for Human Rights Infringement of 2005, submitted
   by the Democratic Party of Japan, and the Bill of the Act for Establishment
   of the Human Rights Council of 2012, submitted by the government.
8 The last report was published in March 2008.
9 See 'Financial Situation of Japan' prepared by the Ministry of Finance in
   December 2015.
10 See the United Nations Budget published on the homepage of the UN.

Chapter 10

1 The international poverty line set by the World Bank was stipulated as $1
   per day in 1990. Later, it was revised to $1.25 in 2008 in line with price
   level changes, then to $1.90 in October 2015.
2 See http://iresearch.worldbank.org/PovcalNet/povDuplicateWB.aspx,
   accessed July 20, 2017.
3 Conversely, there are cases in which income statistics cannot sufficiently
   depict actual conditions. As such, there are instances in which poor
   countries seem poorer than in reality.
4 Factors are complicated, with the gap in scholastic abilities between
   economic classes widening more than that between races (see Reardon 2011).

Chapter 15

1 See http://www.unfoundation.org/.
2 See https://www.unglobalcompact.org/.

3 See https://academicimpact.un.org/.
4 See http://www.leadinggroup.org/.
5 See https://www.unitaid.eu.

Chapter 18

1 See the summarized version in 'L'ABC des Nations Unies', prepared by the UN's Department of Public Information.
2 See the material on the Global Policy Forum's webpage.
3 Accordingly, it outlines the details of each organ: the General Assembly in Chapter IV, the Security Council in Chapter V, the Economic and Social Council in Chapter X, the Trusteeship Council in Chapter XIII, the International Court of Justice in Chapter XIV and the Secretariat in Chapter XV.
4 The official name is 'Act on Cooperation with United Nations Peacekeeping Operations and Other Operations', Act No. 79 of June 19, 1992.

Chapter 19

1 The EU's energy and climate change package set the so-called '20-20-20' targets that have to be achieved by 2020, i.e. a 20% reduction in greenhouse gas emissions from 1990 levels, raising the share of EU energy consumption produced from renewable resources to 20% and a 20% improvement in energy efficiency.
2 For details, please visit the UNDP's website at http://hdr.undp.org/en/content/human-development-index-hdi.
3 For details, please refer to Partha Dasgupta's paper 'Economic development, environmental degradation, and the persistence of deprivation in poor countries' (2002).
4 In brief, the 'weak' sustainability view says that the depletion of natural capital can be compensated by the increase of the sum of manufactured and human capital, whereas according to the 'strong' sustainability perspective, there are certain functions performed by the environment that cannot be duplicated by humans but are crucial for human survival.
5 For details, please visit the UNDP's website at http://www.undp.org/content/undp/en/home/mdgoverview.html.
6 The green transition or transformation is the shift to a green economy that has been taking place at the global, national and local levels.
7 This model has also been known as the 'environmental Kuznets curve'.
8 According to the general consensus in the scientific community, we should prevent an increase in global average temperatures of more than two degrees Celsius above pre-industrial levels by the end of the twenty-first century.
9 The original text is from a 2007 book titled *The Sustainable Development Paradox*, edited by R. Krueger and D. Gibbs.
10 Since 1992 when the Rio Declaration on Environment and Development was adopted.

11 See http://www.degrowth.org/.
12 For example, increasing fuel efficiency makes it cheaper to travel. Therefore, savings from more fuel-efficient cars or planes are outweighed by the rise of travel frequency and the total kilometers traveled. The outcomes are greater fuel consumption and higher GHG emissions.
13 See https://www.water-energy-food.org/.

Chapter 20

1 In this chapter, science refers only to the natural sciences. Similarly, natural scientists are simply referred to as scientists.
2 'Paradigm' is the concept of the 'way of human perception of nature' proposed by science historian Thomas Kuhn (1922–1996) in his book *The Structure of Scientific Revolutions*. Based on 'puzzle solving', i.e. the existing paradigm, and after accepting that paradigm, Kuhn describes the science of trying to deduce answers by questioning nature as 'normal science'. However, he emphasizes that science reaches a crisis situation when a natural phenomenon that cannot be processed is discovered in this paradigm, and it causes a major transformation. He calls this 'paradigm disruption', or in other words, 'scientific revolution'.
3 Abduction is also used as emergence; in other words, it is the transitional equivalent of 'when parts are assembled, properties that are not confined to a simple total of all the properties of the parts appear'. However, in this chapter, it is consistently used as the transitional equivalent of 'abduction'.
4 See http://www.courts.go.jp/app/files/hanrei_jp/178/082178_hanrei.pdf.
5 An ATS had been installed on the train that overturned.

Glossary

Asian Infrastructure Investment Bank (AIIB)
The Asian Infrastructure Investment Bank is a multilateral development bank that aims to support the construction of infrastructure in the Asia-Pacific region. The bank was proposed as an initiative by China in October 2014. The initiative gained support from thirty-seven regional and twenty non-regional Prospective Founding Members. The bank began operating after the agreement entered into force on December 25, 2015. Japan and the US did not join due to concerns regarding the AIIB's governance.

Child poverty rate
This refers to the percentage of children living in poor households.

Citizen scientist
William Whewell (1794–1866), a professor at the University of Cambridge in the UK, proposed the term 'scientist' to describe the profession of a 'natural philosopher'. He thought that 'science' was regarded as distinct and independent from 'philosophy' and other fields of knowledge in the nineteenth century. However, Thomas Henry Huxley (1825–1895), also of the UK, was uncomfortable with the term 'scientist' and wanted to be called a 'man of science', because science, a human activity to discover the unknown, is not a base act to gain financial reward.

In the twentieth century, scientific breakthroughs led to the creation of a variety of new technologies, and a new innovation model called the 'central research laboratory model' was established where the private sector employed scientists to carry out research and development. Consequently, 'scientist' became a professional category at that time. In this book, 'citizen scientist' means 'person of science', which Huxley was thought to have implied. It simply signifies a person who seeks ways to discover the unknown, as distinct from a 'professional scientist'.

Community
It is useful to identify several groups of nodes in terms of the properties of nodes. By considering links as a property of nodes, we identify groups of nodes in order to obtain more links inside each group and less between groups. This way of identifying groups corresponds to maximizing an objective function called 'modularity' defined for a network as a whole. In network science, the groups obtained this way are called 'communities'. One of the most famous studies in this field is an analysis of community formation in a Karate club in the US.

Complex network
Networks consist of nodes and links. When all the nodes that constitute a network have the same degree (K), the network is called a regular network. On the other hand, the degree distribution of a random network is known as a Poisson distribution. A regular network has a uniform degree distribution, whereas a random network has a Poissonian degree distribution. Therefore, a network that falls into an intermediate category between these two types of networks is referred to as a 'small-world network'. Conversely, when the degree of a network exhibits a power-law distribution, the network is called a 'scale-free network'. In a scale-free network, a small number of nodes have very large degrees. Small-world networks and scale-free networks are collectively called 'complex networks'.

Convention Concerning the Protection of the World Cultural and Natural Heritage
The UNESCO General Conference adopted this convention at its seventeenth session in 1972. Registering natural and cultural heritage endowed with 'universal value' for the whole of humanity, its objective is to safeguard this heritage from destruction and damage through the establishment of a system of international cooperation and assistance. The States Parties to this convention have a duty to establish such a system of international assistance regarding the protection, conservation and transmission of world heritage to future generations, including a financial contribution. Japan has adhered to this convention since 1992.

Earth Summit
The UN held the Earth Summit in Rio de Janeiro in 1992. The core issue addressed at the summit was sustainable development with

the aim of environmental conservation. A total of 175 governments participated, with 116 of these sending their heads of state. Some 2,400 representatives of NGOs attended the summit, while 17,000 people attended an NGO 'Global Forum' held in parallel.

Econophysics

Econophysics is a field of science that studies the collective processes inherent in the economy that cross the barrier between the natural and social sciences. In econophysics, data analysis, statistical physics, network science and agent-based modeling play an important role, and it can be said that econophysics marks the beginning of data science. Recently, collaborations between economists and physicists have become more frequent, reflecting the rapid increase of economic data analysis due to the development of information technologies.

Evolutionary-institutional economics

Despite the fact that evolutionary and institutional economics are considered separate fields of thought, they share many common characteristics such as criticism of the assumptions and methods of mainstream economics and emphasis on historical approaches and institutional change or evolution over time. The origins of evolutionary-institutional economics can be found in the American school of institutional economics that developed at the end of the nineteenth and beginning of the twentieth centuries.

Geworfenheit (thrownness), Entwurf (projection)

These concepts were employed as technical terms by Martin Heidegger (1889–1976) in his analysis of the nature of the existence of human beings (*Dasein*) in his most well known work, *Sein und Zeit* (Being and time). *Geworfenheit*, or 'thrownness', refers to being in a situation into which we have been placed through no intention of our own, having been 'thrown' there as something that exists in a particular state of being, and the necessity of dealing with this fact. In contrast to this aspect of our existence, *Entwurf*, or 'projection', refers to not simply being thrown into a certain situation but rather choosing our own manner of being within the constraints and conditions of the situation in which we find ourselves, transcending our present selves in an ongoing process of self-creation. Together, these two aspects of being are referred to as '*geworfener Entwurf*', or 'thrown projection'.

Historical specificity

According to mainstream economics, the behavior of economic actors is governed by the same rules everywhere in the world. Consequently, mainstream economists at the IMF and the World Bank have often recommended the same, one-size-fits-all, economic policies to various developing countries in spite of their differences in historical conditions and local context. No wonder that in most cases these recommendations have proved to be wrong. It can be argued that economic theories and policies should not be universal, i.e. apply everywhere in different historical periods in the same way. On the contrary, economic theories and policies should be tailored to fit the specific historical conditions and local context of the country in question. Adopting such way of thinking also implies taking into account 'historical specificity'.

Human Development Index (HDI)

The United Nations Development Programme (UNDP) has been announcing the Human Development Index (HDI) since the first publication of its Human Development Report in 1990. This index measures the health, education and income levels of each country. In 2010, the HDI was revised and the Multidimensional Poverty Index (MPI) was introduced. From 1997 to 2009, two human poverty indices were used: HPI-I for developing countries and HPI-II for OECD countries.

Human security

Beyond the concept of national security, human security focuses primarily on protecting people suffering from various threats violating peace, and assuring sustainable development. It is needed in response to the complexity of the globalized world. It emphasizes aiding individuals by using a people-centered approach to resolve inequalities that affect security.

Intergovernmental Panel on Climate Change (IPCC)

The IPCC was established in 1988 by the WMO (World Meteorological Organization) and UNEP (United Nations Environment Programme). It aims to provide scientific, technical and socioeconomic evaluations relevant to understanding the scientific basis of risk of human-induced climate change, its potential impacts and options for adaptation and mitigation.

Kategorischer Imperativ (categorical imperative)
In his *Groundwork of the Metaphysics of Morals* (1785) and *Critique of Practical Reason* (1788), Immanuel Kant (1724–1804) writes that the fundamental principle of morality is not an imperative of the form 'if X, do Y' with various conditions attached (a 'hypothetical imperative'), but rather an imperative of the form 'do X' that must be absolutely, unconditionally obeyed (a 'categorical imperative'). Kant formulates this fundamental principle (categorical imperative) as follows: 'only act in accordance with a maxim (rule determining the acts of individuals) that could at the same time be applied universally'.

Mainstream economics
Most economics courses around the world teach mainstream economics. It can be easily recognized by the use of many unrealistic assumptions and mathematical models. Mainstream economics is also the synthesis of neoclassical and Keynesian economics.

Millennium Development Goals (MDGs)
The Millennium Development Goals were established following the Millennium Summit of the United Nations in 2000, after the adoption of the United Nations Millennium Declaration. The MDGs seek to address eight major issues: (1) eradicate extreme poverty and hunger; (2) achieve universal primary education; (3) promote gender equality and empower women; (4) reduce child mortality; (5) improve maternal health; (6) combat HIV/AIDS, malaria and other diseases; (7) ensure environmental sustainability; and (8) develop a global partnership for development. (Source: United Nations Development Program.)

Official Development Assistance (ODA)
Official Development Assistance is defined as those flows to countries and territories on the DAC (Development Assistance Committee) List of ODA recipients and to multilateral development institutions, which are:
1. provided by official agencies, including state and local governments, or by their executive agencies; and

2. each transaction of which:
    a) is administered with the promotion of the economic
       development and welfare of developing countries as its
       main objective; and
    b) is concessional in character and conveys a grant element of
       at least twenty-five percent (calculated at a rate of discount
       of ten percent).
    (Source: Organization for Economic Co-operation and
    Development.)

Output fluctuation

Recently, much attention has been paid to renewable energy due to
its inherent property of zero greenhouse gas emissions. Renewable
energy is categorized as dispatchable, such as hydro-power and
biomass power or non-dispatchable, such as solar photo-voltaic
power and wind power. Here, 'dispatchable energy' means its
output power can be controlled. However, non-dispatchable energy
cannot be controlled and its output power fluctuates over a short
period of time, depending on the weather conditions. The stable
operation of a power grid system requires a supply/demand balance
over a short period of time. For this reason, a system operator has
to compensate for the output fluctuation from the non-dispatchable
renewable energy using thermal power plants and electric storage.
In a conventional power grid, it is considered difficult in practice to
integrate non-dispatchable renewable energy at a level greater than
twenty percent of the total electricity of the grid system. Therefore,
we need technological innovations, such as low cost electric storage,
demand-side management and smart inverter systems, etc.

Participatory development

Participatory development is a process through which stakeholders
can influence and share control over development initiatives, and
over the decisions and resources that affect them (source: Asian
Development Bank). Through this approach to development, the
greatest number of people possible can be involved in development
decision-making and enjoy its benefits.

Path dependence and lock-in

The concept of path dependence is often used to explain the
variety of outcomes arising from the post-socialist transition to

a market economy. For instance, the progress towards a market economy in the former socialist countries in South-Eastern Europe and the former USSR, excluding the three Baltic countries, was slower and much more difficult in comparison with that of the Central-Eastern European countries. The main reason was the different historical conditions in those groups of countries at the time when the socialist system collapsed. The more unfavorable initial conditions eventually led to a different transition trajectory in South-Eastern Europe and the former USSR, characterized by the lock-in to a 'bad equilibrium' of corruption, state capture and organized crime.

Poverty rate
There are two measures of 'poverty rate' based on either income or consumption: the absolute poverty rate and the relative poverty rate. Since 1990, the World Bank has defined the absolute poverty rate as the percentage of the population living below the threshold of one dollar per day; this amount was raised to $1.25 in 2008 and $1.90 in 2015. Some countries announce their absolute poverty rate based on a certain income or consumption threshold. The OECD uses the percentage of individuals in households with an equivalized disposable income below half the median. This is called the 'relative poverty rate'.

Radiocarbon dating
Radiocarbon dating is a method for determining the age of organic materials by using the properties of radiocarbon (^{14}C).

Stable isotope analysis
Isotope analysis has widespread applicability in many fields of sciences. These include numerous applications in the biological and environmental sciences, and also in archaeology. ^{18}O is used for paleoclimatology, ^{13}C and ^{15}N are used for reconstructing the paleo-diet.

Sustainable Development Goals (SDGs)
At the United Nations Sustainable Development Summit on September 25, 2015, the target year of the MDGs, more than 150 world leaders adopted the new 2030 Agenda for Sustainable Development, including the Sustainable Development Goals (SDGs).

The seventeen new Sustainable Development Goals aim to end poverty, hunger and inequality, take action on climate change and the environment, improve access to health and education, build strong institutions and partnerships, and more. (Source: United Nations Development Program.)

Synchronization
In 1673, the Dutch physicist C. Huygens placed two pendulum clocks with slightly different periods of oscillation on different posts, showing that with different periods of oscillation the time displayed by the two clocks gradually differs. However, when these two pendulum clocks were attached to the same beam, he discovered that they performed a complete swing in exactly the same period of time – the two pendulum clocks placed on the same beam continued to beat the same time. When there is an interaction between two pendulums, they begin to swing in the same period with fixed phases. This phenomenon is called synchronization. Certain species of fireflies that live in Southeast Asia are known for their habit of flocking together on a single tree with their lights blinking in unison.

Transilience of knowledge
In issue-oriented research, the problem to be solved always requires transdisciplinary knowledge. For instance, quantum mechanical knowledge, innovation theory and industry theory are all necessary in the study of competitiveness for the semiconductor/ nanotechnology industries. Furthermore, knowledge of corporate governance and nuclear/high pressure physics are needed simultaneously to identify the root causes of the Fukushima nuclear power plant accident. In this way, understanding the essence of contemporary issues requires crossing borders concurrently, and the practice of accomplishing this is called 'transilience of knowledge'. The concept 'transilience' was originally proposed by sociologist Anthony Richmond in 1969 to represent the activity of crossing national borders in post-industrial societies.

United Nations Millennium Declaration
In September 2000, following a three-day Millennium Summit of world leaders involving 149 heads of state and government and high-ranking officials from over forty other countries at the headquarters of the United Nations, the General Assembly adopted

the Millennium Declaration committing their nations to a new global partnership to reduce extreme poverty and setting out a series of time-bound targets, with a deadline of 2015, that have become known as the Millennium Development Goals. (Source: United Nations Development Program.)

United Nations Educational, Scientific and Cultural Organization (UNESCO)

UNESCO was established at the United Nations in 1946 on the basis of its charter adopted in 1945. According to the definition given in this charter, it was established with the goal of international peace and common welfare. In this aim, it has pursued major objectives such as 'education for all', 'the preservation of cultural diversity' and 'the promotion of dialogue among civilizations'. It has surveyed the progress of literacy rates, measured the diffusion of compulsory education, advanced the registration and protection of world heritage and fostered the preservation of cultural diversity.

United Nations Millennium Summit

This summit was held in New York City in 2000, focusing on various global issues such as poverty, AIDS and the equitable distribution of the benefits that arise with globalization. One hundred and eighty-nine member states of the UN agreed to help citizens in the world's poorest countries achieve a better life by 2015. The Millennium Development Goals (MDGs) were set and the framework and progress outline was established.

World Summit on Sustainable Development

The UN held this summit in Johannesburg in 2002, ten years after the Earth Summit in Rio de Janeiro, and many business leaders and NGOs participated.

Bibliography

Acemoğlu, D. and J.A. Robinson (2012) *Why Nations Fail: The Origins of Power, Prosperity and Poverty.* New York: Crown Business.

AFP News (2014) Big data becomes tool in Ebola battle. Accessed on September 1, 2017 at http://medicalxpress.com/news/2014-10-big-tool-ebola.html.

Airapetian, V.S. et al. (2016) Prebiotic chemistry and atmospheric warming of early Earth by an active young Sun. *Nature Geoscience,* 9: 452–455.

Akasaka, K. (2014) *Kokusai Kikan de Mita 'Sekai no Elite' no Shōtai* (The identities of the world's elite that I observed in international organizations). Tokyo: Chūkō Shinsho Rakure.

Akimoto, Y. (1995) *Shinayakana Seiki* (Flexible century). Tokyo: Japan Electric Association, Newspaper Division.

Allison, G. and P. Zelikow (1999) *Essence of Decision: Explaining the Cuban Missile Crisis,* second edition. New York: Longman.

Allison, G.T. (1971) *Essence of Decision: Explaining the Cuban Missile Crisis.* Boston: Little, Brown & Co.

Angrist, J.D. and J.S. Pischke (2009) *Mostly Harmless Econometrics: An Empiricist's Companion.* Princeton: Princeton University Press.

Angrist, J.D. and V. Lavy (1999) Using Maimonides' Rule to estimate the effect of class size on scholastic achievement. *The Quarterly Journal of Economics,* 114(2): 577–599.

Annan, K.A. (2000) *We the Peoples: The Role of the United Nations in the 21st Century.* New York: United Nations Department of Public Information.

Aoki, M. (2001) *Toward a Comparative Institutional Analysis.* Cambridge: MIT Press.

AP News (2015) Searching for ET: Hawking to look for extraterrestrial life. Accessed on September 1, 2017 at http://www.chicagotribune.com/news/nationworld/ct-stephen-hawking-extraterrestrial-life-20150720-story.html.

Arjava, A. (2005) The mystery cloud of 536 CE in the Mediterranean sources. *Dumbarton Oaks Papers,* 59: 72–94.

Arthur, W.B. (1994) Inductive reasoning and bounded rationality (The El Farol Problem). *American Economic Review,* 84: 406.

Ashok, K. and T. Yamagata (2009) The El Nino with a difference. *Nature,* 461: 481–484.

Ashok, K., S.K. Behera, S.A. Rao, H. Weng and T. Yamagata (2007) El Nino Modoki and its possible teleconnection. *Journal of Geophysics Research,* 112(C11007).

Azuma, H. et al. (2011) *IT Jidai no Shinsai to Kaku Higai* (Earthquake disaster and nuclear accident in the IT era). Tokyo: The Impress Press.

Baba, H. (ed.) (2012) Special edition: Latest trend of paleo-anthropological studies. *Quarterly Archaeology,* 118: 4–99.

Banerjee, A., A.V. Banerjee and E. Duflo (2011) *Poor Economics: A Radical*

Rethinking of the Way to Fight Global Poverty. New York: Public Affairs.

Banzai, H. (2006) *Kokusai Kikō to wa Nani ka* (What are international organizations). Tokyo: Iwanami Shoten.

Barabási, A.-L. and E. Bonabeau (2003) Scale-free networks. *Scientific American*, 288: 50–59.

Batalha, N.M. et al. (2013) Planetary candidates observed by Kepler III. Analysis of the first 16 months of data. *The Astronomical Journal Supplement Series*, 204(2): 24.

Bergmann, J. and A. Sams (2012) *Flip Your Classroom: Reach Every Student in Every Class Every Day*. Arlington: International Society for Technology in Education.

Bergson, H. (1959) L'énergie spirituelle (Spiritual energy). In H. Bergson, *Œuvres* (Works). Textes annotés par André Robinet. Paris : PUF.

Bill and Melinda Gates Foundation (2017) *What We Do*. Accessed on September 1, 2017 at http://www.gatesfoundation.org/What-We-Do.

Brady, H.E. and D. Collier (eds) (2010) *Rethinking Social Inquiry: Diverse Tools, Shared Standards*, second edition. Lanham: Rowman & Littlefield.

Braudel, F. (2004 [1949]) *Braudel Rekishi Shūsei 1 Chichukai*. Japanese translation of *La Mediterranee et le monde mediterraneen a l'époque de Philippe II, Librairie Armand Colin, 2ᵉ edition revue et corrigée*, trans. Y. Hamana. Tokyo: Fujiwara Shoten Publishers.

Burroughs, W.J. (2005) *Climate Change in Prehistory: The End of the Reign of Chaos*. Cambridge: Cambridge University Press.

Cabinet Office (2007) *Hanshin Awaji Daishinsai Kyōkun Jōhō Shiryōshū* (Lessons of the Great Hanshin-Awaji Earthquake). Accessed on September 1, 2017 at http://www.bousai.go.jp/kyoiku/kyokun/ hanshin_awaji/index.html.

Campbell, D.T. and D.W. Fiske (1959) Convergent and discriminant validation by the multitrait-multimethod matrix. *Psychological Bulletin*, 56(2): 81–105.

Cann, R.L., M. Stoneking and A.C. Wilson (1987) Mitochondrial DNA and human evolution. *Nature*, 325: 31–36.

Carson, R. (1962) *Silent Spring*. Boston: Houghton Mifflin.

Childe, G. (1951 [1936]) *Bunmei no Kigen*. Japanese translation of *Man Makes Himself*, trans. M. Nezu. Tokyo: Iwanami Shoten Publishers.

Christian, D. (2004) *Maps of Time: An Introduction to Big History*. Berkeley: University of California Press.

Cohen, M.D., J.G. March and J.P. Olsen (1972) A garbage can model of organizational choice. *Administrative Science Quarterly*, 1–25.

Coleman, J.S. (1968) The concept of equality of educational opportunity. *Harvard Educational Review*, 38(1): 7–22.

Council on Foreign Relations (2012) Are randomized control trials a good way to evaluate development projects?. August 10, 2012. Accessed on September 7, 2017 at https://www.cfr.org/blog/question-week-are-randomized-controlled-trials-good-way-evaluate-development-projects.

Dasgupta, P. (2002) Economic development, environmental degradation, and the persistence of deprivation in poor countries. A background paper for the United Nations' *World Summit on Sustainable Development,* August 26 to September 4, 2002, Johannesburg. Accessed on September 7, 2017 at http://faculty.cbpp.uaa.alaska.edu/elhowe/ECON_F04/ dasgupta_wb_02.pdf.

Descartes, R. (2008 [1637]) *Discourse on the Method.* Oxford: Oxford University Press.

Diamond, J. (2000) *Jū, Byōgenkin, Tetsu.* Japanese translation of *Guns, Germs, and Steel: The Fates of Human Societies,* trans. A. Kurahone. Tokyo: Sōshisha.

Dōgen (1993) *Shōbōgenzō* (Treasury of the true dharma eye). Tokyo: Iwanami Bunko.

Dror, Y. (1983) *Public Policymaking Reexamined,* second edition. New Brunswick: Transaction Publishers.

Esaki, R. (1995) *Saiensu no Shinzui* ('Essence of science' structure of living things). *Kōzō Seibutsu* (Structure biology), 1(1).

Fishkin, J.S. (2009) *When the People Speak: Deliberative Democracy and Public Consultation.* Oxford: Oxford University Press.

Food and Agriculture Organization of the United Nations (FAO) (2011) *The State of the World's Forests 2011.* Rome: FAO.

Fujita, M., D.M. Prawiradilaga and T. Yoshimura (2014) Role of fragmented and logged forests for bird communities in industrial *Acacia mangium* plantations in Indonesia. *Ecological Research,* 29(4): 741–755.

Galbraith, J.K. (1958) *The Affluent Society.* Boston: Houghton Mifflin.

Gamoran, A. and D.A. Long (2007) Equality of educational opportunity, a 40 year retrospective. In R. Teese, S. Lamb and M. Duru-Bellat (eds) *International Studies in Educational Inequality, Theory and Policy Volume 1: Educational Inequality: Persistence and Change.* Dordrecht: Springer, pp. 23–47.

Gartner (2011) Gartner says solving 'big data' challenge involves more than just managing volumes of data. Accessed on September 1, 2017 at http:// www.gartner.com/newsroom/id/1731916.

Gates, B. and M. (2015) *Our Big Bet for the Future* (annual letter). Accessed on September 1, 2017 at http://www.gatesnotes.com/2015-Annual-Letter.

Gell-Mann, M. (1995) *The Quark and the Jaguar,* third edition. New York: Griffin.

George, A.L. and A. Bennett (2005) *Case Studies and Theory Development in the Social Sciences.* Cambridge: MIT Press.

Greenhalgh, T. and B. Hurwitz (1998) *Narrative Based Medicine: Dialogue and Discourse in Clinical Practice.* London: BMJ Books.

Gunn, J.D. (2000) *The Years Without Summer: Tracing A.D. 536 and Its Aftermath.* BAR International Series 872. Oxford: Archaeopress.

Haber, J. (2014) *MOOCs.* Cambridge: MIT Press.

Haken, H. (1977) *Synergetics: An Introduction: Nonequilibrium Phase Transitions and Self-Organization in Physics, Chemistry and Biology.* New York: Springer.

Hara, T. (2010) Shinrin no Bussitsu Seisan (Matter production of forests). In Research Institute for Humanity and Nature (ed.) *Chikyū Kankyōgaku*

Jiten (Encyclopedia of global environmental studies). Kyoto: Kobunsha, pp. 50–51.

Harrington, M. (1962) *The Other America: Poverty in the United States*. New York: Macmillan.

Hasumi, T. (2006) Tsūshō Kin'yū to Ahakaimondai: Gurōbaru-ka to Kokusai Kikō (Trade, finance and social problems: Globalization and international organizations). In S. Katsuhiro (ed.) *Kokusai Kikan* (International organizations). Tokyo: Iwanami Shoten.

Hawking, S. (2014) Hawking warns on rise of the machines. *Financial Times*. Accessed on September 1, 2017 at https://www.ft.com/content/9943bee8-7a25-11e4-8958-00144feabdc0.

Hayami, A. (2009) *Nihon o Osotta Supein Infuruenza* (Spanish Influenza that ravaged Japan). Tokyo: Fujiwara Shoten.

Hayami, A. (2012) *Rekishi Jinkōgaku no Sekai* (The world of historical demography). Tokyo: Iwanami Shoten.

Hayashi, C., K. Nakazawa and Y. Nakagawa (1985) Kyoto Model. In D.C. Black and M.S. Matthews (eds) *Protostars and Planets II*. Tucson: The University of Arizona Press.

Haynes, L., O. Service, B. Goldacre and D. Torgerson (2012) *Test, Learn, Adapt: Developing Public Policy with Randomized Controlled Trials*. London: Cabinet Office Behavioural Insights Team. Accessed on August 9, 2017 at http://www.gov.uk/government/uploads/system/uploads/attachment_data/file/62529/TLA-1906126.pdf.

Heckman, J.J. (2006) Skill formation and the economics of investing in disadvantaged children. *Science*, 312(5782): 1900–1902.

Heidegger, M. (1972) *Sein und Zeit* (Being and time), twelfth edition. Tübingen: Niemeyer.

Hirano, T. (2013) Higashi Nihon Daishinsai ni okeru Denryokugaisha no Risuku Taiō Hyōka no Saikentō: 'Rekishi Tsunami' ni taisuru Risuku Taiō no Jirei Bunseki o tsūjite (Re-investigation of the assessment of risk response carried out by the power companies during the Great East Japan Earthquake: Through the analysis of a risk response case-study carried out on the 'historical tsunami'). *Japan Business Ethics Journal*, 21: 71–85.

Hodgson, G. (2001) *How Economics Forgot History: The Problem of Historical Specificity in Social Science*. London: Routledge.

Hoffmann, U. (2014) The technical and socio-economic pitfalls of green growth: A reality check to avoid disillusionment. Paper presented at the *International Symposium on Green Growth and Global Environmental Change*, Institute for the Advanced Study of Sustainability, UN University, July 25–26, 2014, Tokyo.

Holland, J. (1996) *Hidden Order: How Adaptation Builds Complexity*. Reading: Helix Books.

Hubble, E. (2013 [1936]) *The Realm of the Nebulae*. New Haven: Yale University Press.

Huygens, C. (1986 [1673]) *The Pendulum Clock or Geometrical Demonstrations Concerning the Motion of Pendula As Applied to Clocks*. Iowa: Iowa State University Press.

Ikejima, Y. (2014) *Kokusai Kikan no Seiji Keizai-gaku* (Political economy of international organizations). Kyoto: Kyoto University Press.

IPCC (2014) *Summary for Policy Makers*. Accessed on September 1, 2015 at http://report.mitigation2014.org/spm/ipcc_wg3_ar5_summary-for-policymakers_approved.pdf.

Japan Forestry Agency (2001) *Shinrin Ringyō Kihon Keikaku* (Basic Plan for Forest and Forestry). Tokyo: Japan Forestry Agency.

Japan Oil, Gas and Metals National Corporation (JOGMEC) (2014) *Mineral Resource Material Flows*. Japan Oil, Gas and Metals National Corporation.

Jonas, H. (1979) *Das Prinzip Verantwortung* (The imperative of responsibility). Frankfurt am Main: Insel-Verlag.

Jonas, H. (1984 [1979]) *The Imperative of Responsibility*. Translated by H. Jonas with D. Herr. Chicago: University of Chicago Press.

Kaji, M. (2014) *Kokusai Shakai de Hataraku Kokuren no Genba kara Mieru Sekai* (The world seen through the United Nations, working amidst the international community). Tokyo: NTT publishing.

Kamis, A. and D. Taylor (eds) (1993) Acacia mangium *Growing and Utilization*. Bangkok: Winrock International and The Food and Agriculture Organization of the United Nations.

Kamo no Chōmei (1957) *Hōjōki* (An account of my hut). *Nihon koten bungaku taikei* (An anthology of Japanese classical literature) *Vol. 30 'Hōjōki/ Tsurezuregusa'*. Tokyo: Iwanami Shoten.

Karaki, J. (1964) *Mujō* (Transience). Tokyo: Chikuma Shobō.

Kato, S. (2013) *Jinrui to Kansenshō no Rekishi: Michi naru Kyōfu o Koete* (History of humans and infections: Overcoming fear of the unknown). Tokyo: Maruzen.

Kawai, S. and K. Watanabe (2016) Biomass assessment of afforested regions in tropical Southeast Asia. In K. Mizuno, M. Fujita and S. Kawai (eds) *Catastrophe & Regeneration in Indonesia's Peatlands: Ecology, Economy & Society*. Singapore: NUS Press, pp. 93–115.

Kawashima, H. (2015) Gareki no Shita no Iryō, IT no Yakuwari wa? (Medical services in a disaster situation: What is the role of IT?). *Cloud Watch*. Accessed on September 1, 2017 at http://cloud.watch.impress.co.jp/docs/case/20150117_681906.html.

Keating, J. (2014) Random acts: What happens when you approach global poverty as a science experiment?. March 26, 2014. Accessed on September 7, 2017 at http://www.slate.com/articles/business/crosspollination/2014/03/randomized_controlled_trials_do_they_work_for_economic_development.html.

Kenshi, K. (2006) *Posuto Kyōtogiteisho?: Kikō Hendō o Meguru Chōki-Teki Kokusai Seido Giron no Yukue: Kokuren to Chikyū Shimin Shakai no Atarashii Chihe* (Post-Kyoto Protocol: The whereabouts of long-term international institutional debate over climate change, new horizons of the United Nations and global civil society). Tokyo: Tōshindō.

King, G., R.O. Keohane and S. Verba (1994) *Designing Social Inquiry: Scientific Inference in Qualitative Research*. Princeton: Princeton University Press.

Kingdon, J.W. and J.A. Thurber (1984) *Agendas, Alternatives, and Public Policies*. Boston: Little, Brown.

Kobayashi, M. (2006) *Takoku-kan Kankyōjōyaku Purosesu no hi Kokka Akutā no Sanka* (The participation of non-state actors in the multilateral environmental treaty process). *Kokuren to chikyū shimin shakai no atarashī chihei'*. Tokyo: Tōshindō.

Kobayashi, S., Y. Omura, K. Sanga-Ngoie, Y. Yamaguchi, R. Widyorini, M. Fujita, B. Supriadi and S. Kawai (2015) Yearly variation of acacia plantation forests obtained by polarimetric analysis of ALOS PALSAR data. *IEEE J of Selected Topics in Applied Earth Observations and Remote Sensing*, 8(11): 5294–5304.

Kobayashi, T. (2007) *Toransu Saiensu no Jidai* (The age of trans-science). Tokyo: NTT Publishing Co.

Kokusai Rengō Kōhō-kyoku (2011) *Kokusai Rengō no Kiso Chishiki* (Basic facts about the United Nations). Translated by M. Hachimori. Hyōgo: Kanseigakuin Daigaku Shuppan-kai.

Konishi, S. (2013) *Enerugī Mondai no Gokai: Ima Sore o Toku* (Misunderstandings of energy issues and resolutions). Tokyo: Kagaku-Dojin.

Konoe, T. (2006) *Shimin Shakai to Saigai Kyūen: Kokuren to Chikyū Shimin Shakai no Atarashii Chihei* (Civil society and disaster relief: New horizons of the United Nations and global civil society). Tokyo: Tōshindō.

Kopparapu, R.K. et al. (2013) Habitable zones around main-sequence stars: New estimates. *The Astrophysical Journal*, 765(2): 131.

Kouzai, S. (1991) *Kokuren no Heiwa Iji Katsudō* (Peace keeping operations of the United Nations). Tokyo: Yūhikaku.

Koyasu, S., A. Nomoto and M. Mitsuyama (eds) (2011) *Men'eki Kansen Seibutsugaku (Gendai Seibutsu Kagaku Nyūmon 5)* (Immuno and infection biology: Introduction to contemporary biology, volume 5). Tokyo: Iwanami Shoten.

Krugman, P. (2012) Economics in the Crisis. *New York Times*, March 5, 2012. Accessed on September 7, 2017 at http://krugman.blogs.nytimes.com/2012/03/05/economics-in-the-crisis/.

Kubota, J. (ed.) (2012) *Chūō Eurasia Kankyōshi* (Environmental history of Central Asia). Tokyo: Rinsen Shoten Publishers.

Kudo, Y. (2012) *Kyūsekki Jōmon Jidai no Kankyō Bunkashi* (History of cultural environment in Paleolithic and Jomon Age). Tokyo: Shinsen Sha Publishers.

Kuhn, T. (1971) *Kagaku Kakumei no Kōzō*. Japanese translation of *The Structure of Scientific Revolutions*, trans. S. Nakayama. Tokyo: Misuzu Shobō.

Kuramoto, Y. (2003 [1984]) *Chemical Oscillations, Waves, and Turbulence*. Mineola: Dover Books on Chemistry.

Larsen, L.B. et al. (2008) New ice core evidence for a volcanic cause of the A.D. 536 dust veil. *Geophysical Research Letters*, 35(4708): 1–5.

Lasswell, H.D. (1951) The policy orientation. In D. Lerner and H.D. Lasswell (eds) *The Policy Sciences: Recent Developments in Scope and Methods*. Stanford: Stanford University Press.

Lasswell, H.D. (1971) *A Pre-View of Policy Science*. New York: American Elsevier.

Levine, P.B. and D.J. Zimmerman (eds) (2010) *Targeting Investments in Children: Fighting Poverty when Resources are Limited*. Chicago: University of Chicago Press.

Lewis, O. (1959) *Five Families: Mexican Case Studies in the Culture of Poverty*. New York: Basic Books.

Lidskog, R. and I. Elander (2012) Ecological modernization in practice? The case of sustainable development in Sweden. *Journal of Environmental Policy and Planning*, 14(4): 411–427.

Lindblom, C.E. and E.J. Woodhouse (1968) *The Policy-Making Process*. Englewood Cliffs: Prentice-Hall.

Linz, B. and F. Balloux et al. (2007) African origin for the intimate association between humans and *Helicobacter pylori*. *Nature*, 445: 915–918.

Lipsey, M.W. et al. (2013) *Evaluation of the Tennessee Voluntary Pre-kindergarten Program: Kindergarten and First Grade Follow-Up Results from the Randomized Control Design*. Nashville: Peabody Research Institute, Vanderbilt University.

Lorek, S. and J. Spangenberg (2014) Sustainable consumption within a sustainable economy: Beyond green growth and green economies. *Journal of Cleaner Production*, 63: 33–44.

Lorenz, E.N. (1963) Deterministic nonperiodic flow. *Journal of the Atmospheric Sciences*, 20(2): 130–141.

Maehara, H. et al. (2012) Superflares on solar-type stars. *Nature*, 485: 478–481.

Mann, H. (2010 [1848]) Twelfth annual report of the secretary of the Massachusetts school board. In A.J. Milson et al. (eds) *American Educational Thought: Essays from 1640-1940*. Charlotte: Information Age Publishing, p. 168.

Matuura, K. (2011) *Kokusaijin no Susume* (An internationalist's recommendations). Tokyo: Seizansha.

Mayor, M. and D. Queloz (1995) A Jupiter-mass companion to a solar-type star. *Nature*, 378(23): 355–359.

Meadows, D.H. et al. (1972) *The Limits to Growth*. New York: Universe Books.

Miichi, M. (1994) *Korera no Sekaishi* (World history of cholera). Tokyo: ShobunSha.

Miki, K. (1966) *Miki Kiyoshi Zenshū* (Complete works of Miki Kiyoshi). Vol. 1, Tokyo: Iwanami Shoten.

Mikoto, T. (2006) *Kokuren to Bijinesu no Pātonāshippu: Kokuren to Chikyū Shimin Shakai No Atarashī Chihei* (United Nations and the business of partnership: New horizons of the United Nations and global civil society). Tokyo: Tōshindō.

Mitsubishi UFJ Research & Consulting (2014) Recent trends in policy analysis and estimation of the effect of repeated requests for response to a questionnaire survey by using RCT. Policy Research Report, October 10, 2014. Accessed on September 7, 2017 at http://www.murc.jp/thinktank/rc/politics/politics_detail/seiken_141010.pdf.

Miyahara, H. (2008) Cosmic ray & solar activity. *KAGAKU*, 79(12): 1380–1382.

Mizushima, T. (2010) *Gurōbaru Hisutorī Nyūmon (Sekaishi Riburetto 127)* (Handbook of global history, libretto of world history 127). Tokyo: Yamakawa Shuppan Publishers.

Modernising Government Secretariat, Cabinet Office (1999) *Modernising Government*. London: The Stationery Office. Accessed on August 9, 2017 at http://webarchive.nationalarchives.gov.uk/20080609160619/ http://archive.cabinetoffice.gov.uk/moderngov/whtpaper/index.htm.

Mogami, T. (2006) *Kokusai Kikō Ron* (International organizations). Tokyo: University of Tokyo Press.

Mol, A. and D. Sonnenfeld (2000) Ecological modernization around the world: An introduction. *Environmental Politics*, 9(1): 1–14.

Mukai, H. (2012) *Mori to Umi o Musubu Kawa* (Rivers linked with forests and sea). Field Science Education and Research Center, Kyoto University (ed.). Kyoto: Kyoto University Press.

Murase, S. and M. Mayama (2004) *Buryoku Hunsō no Kokusaihō* (International laws on armed conflicts). Tokyo: Tōshindō.

Nakanishi, H., J. Ishida and M. Tadokoro (2013) *Kokusai Seiji Gaku* (International politics: Theories and perspectives). Tokyo: Yūhikaku.

Nakayama, T. (2013) *Genso Senryaku* (Strategic use of elements). Tokyo: Diamond.

NASA News (2013) At least one in six stars has an earth-sized planet. Published January 10, 2013, accessed October 21, 2014 at http://www.nasa.gov/mission_pages/kepler/news/17-percent-of-stars-have-earth-size-planets.html.

Neumann, J.V. and O. Morgenstern (2007 [1944]) *Theory of Games and Economic Behavior*. Princeton: Princeton University Press.

New, M., D. Liverman, H. Schroder and K. Anderson (2011) Four degrees and beyond: The potential for a global temperature increase of four degrees and its implications. *Philosophical Transactions of the Royal Society A*, 369: 6–19.

Nicolis, G. and I. Prigogine (1977) *Self-Organization in Nonequilibrium Systems: From Dissipative Structures to Order through Fluctuations*. Boston: Wiley.

North, D. (1990) *Institutions, Institutional Change and Economic Performance*. Cambridge: Cambridge University Press.

Oakley, A. (2000) *Experiments in Knowing: Gender and Method in the Social Sciences*. New York: The New Press.

OECD (2001) *Starting Strong: Early Childhood Education and Care*. Paris: OECD Publishing.

OECD (2006) *Starting Strong II: Early Childhood Education and Care*. Paris: OECD Publishing.

OECD (2011a) *Divided We Stand: Why Inequality Keeps Rising*. Paris: OECD Publishing.

OECD (2011b) *Towards Green Growth*. Paris: OECD Publishing.

OECD (2012) *Starting Strong III: A Quality Toolbox for Early Childhood Education and Care*. Paris: OECD Publishing.

OECD (2013) *Putting Green Growth at the Heart of Development*. Paris: OECD Publishing.

OECD (2015a) *Starting Strong IV: Monitoring Quality in Early Childhood Education and Care*. Paris: OECD Publishing.

OECD (2015b) *In It Together: Why Less Inequality Benefits All*. Paris: OECD Publishing.

Ohkawara, K. (2012) Higashi Nihon Daishinsai kara Ichinen: IT Sangyo ha Donna Koken ga Dekita no ka? (What contributions could the IT industry make?: One year after the Great East Japan Earthquake). *PC Watch*. Accessed on September 1, 2017 at http://pc.watch.impress.co.jp/docs/column/gyokai/20120312_518131.html.

Ohto, C. (2012) *Rekishi to Jijitsu* (History and fact: Beyond the criticisms of history in post-modernism). Kyoto: Kyoto University Press.

Ono, A. (2000) Shizenshi to Jinruishi (Natural history and human history). In *Kankyō to Jinrui: Shizen no naka ni Rekishi o Yomu* (Environment and human beings, reading the history in nature). Tokyo: Asakura Shoten Publishers, pp. 2–10.

Parker, G. (2013) *Global Crisis: War, Climate Change and Catastrophe in the Seventeenth Century*. New Haven: Yale University Press.

Pascal, B. (1962) *Pensées* (Thoughts). Texte établi par Louis Lafuma. Paris: Éditions du Seuil.

Pierce, C.S. (1965) *Collected Papers of Charles Sanders Peirce, Volume 5*. Cambridge: Belknap Press.

Piketty, T. (2014) *Capital in the Twenty-First Century*. Cambridge: Belknap Press.

Polanyi, M. (1962) The republic of science: Its political and economic theory. *Minerva*, 1: 54–73.

Popova, O.P. and P. Jenniskens et al. (2013) Chelyabinsk Airburst, Damage Assessment, Meteorite Recovery, and Characterization. *Science*, 342(6162): 1069–1073.

Ratnam J.V., S.K. Behera, Y. Masumoto, K. Takahashi and T. Yamagata (2010) Pacific Ocean origin for the 2009 Indian summer monsoon failure. *Geophysical Research Letters*, 37(L07807).

Reardon, S.F. (2011) The widening academic achievement gap between the rich and the poor: New evidence and possible explanations. In G.J. Duncan and R.J. Murnane (eds) *Whither Opportunity? Rising Inequality, Schools, and Children's Life Chances*. New York: Russell Sage Foundation.

REEEP (2014) *Making the Case: How Agrifood Firms Are Building New Business Cases in the Water-Energy-Food Nexus*. Renewable Energy and Energy Efficiency Partnership (REEEP) and the Food and Agriculture Organization of the United Nations (FAO). Accessed on September 7, 2017 at https://www.reeep.org/making-case-how-agrifood-firms-are-building-new-business-cases-water-energy-food-nexus.

Resnick, D., F. Tarp and J. Thurlow (2012) The political economy of green growth: Cases from Southern Africa. *Public Administration and Development*, 32: 215–228.

Sajima, N. (2004) *Gendai Anzen Hoshō Yōgo Jiten* (Contemporary dictionary of security terms). Tokyo: Shinzansha Publisher Co., Ltd.

Sakaki, Y., J. Yamagiwa, N. Arai and O. Karatsu (2014) *Ningen towa Nanika* (What is human being). Tokyo: Kagaku Dojin.

Sangawa, A. (1992) *Jishin Kōkogaku: Iseki ga Kataru Jishin no Rekishi* (Archaeology of earthquakes: History of earthquakes in archaeological sites). Tokyo: Chuokoron Sha Publishers.

Sartre, J.-P. (2007 [1946]) *Existentialism is a Humanism*. Translated by C. Macomber. New Haven: Yale University Press.

Sasaki, S. (2007) Foreword: The role of forest in biosphere. In S. Sasaki, Y. Kihira and K. Suzuki (ed.) *Forest Science*. Tokyo: Bun-eido, pp. 1–15.

Sasao, H. (2000) *Enerugī, Shizen, Chiikishajkai* (Energy, nature and local communities). Tokyo: ERC Publishing Company.

Satō, I. (2006) *Fieldwork: Sho o Motte Machi e Deyo* (Fieldwork: Let us join the crowds in town while carrying books in hands), revised edition. Tokyo: Shinyosha.

Satō, T. (2005) *Kokusai Soshikihō* (International organizations law). Tokyo: Yūhikaku.

Sawada, Y. (1981) A thermodynamic variational principle in nonlinear non-equilibrium phenomena. *Progress of Theoretical Physics*, 66: 68–76.

Schrödinger, E. (1944) What is life?. In E. Schrödinger, *The Physical Aspect of the Living Cell*. Cambridge: Cambridge University Press.

Schrödinger, E. (2012 [1944]) *What is Life?: With Mind and Matter*. Cambridge: Cambridge University Press.

Schultz, T.W. (1961) Investment in human capital. *The American Economic Review*, 51(1): 1–17.

Schweinhart, L.J. et al. (2005) *Lifetime Effects: The High/Scope Perry Preschool Study Through Age 40*. Ypsilanti: High/Scope Press.

Scott, R.A. and A. Shore (1979) *Why Sociology Does Not Apply: A Study of the Use of Sociology in Public Policy*. New York: Elsevier.

Sen, A. (1991) Public action to remedy hunger. *Interdisciplinary Science Reviews*, 16(4): 324–336.

Sen, A. (1992) *Inequality Reexamined*. Cambridge: Harvard University Press.

Sen, A. (1999) *Development as Freedom*. Oxford: Oxford University Press.

Sen, A. (2002) Basic education and human security. A background paper for the *Basic Education and Human Security* workshop, jointly organized by the Commission on Human Security, UNICEF, the Pratichi (India) Trust and Harvard University, and held in Kolkata, January 2–4, 2002.

Shibatani, A. (1973) *Han Kagakuron* (Anti-science theory). Tokyo: Misuzu Shobō, pp. 164–169.

Shimizu, B. (2004) *Kansenshō to dō Tatakau ka* (How to fight infections). Tokyo: Tokyo Kagaku Dōjin.

Simon, H.A. (1996 [1969]) *The Sciences of the Artificial*, third edition. Cambridge: MIT Press.

Snyder, T.D. et al. (2016) *Digest of Education Statistics 2015* (NCES 2016-014). Washington, DC: National Center for Education Statistics, Institute of Education Sciences, U.S. Department of Education.

Soeda, Y. et al. (2012) Paleo-Environment of Paleo-Ainu Period reconstructed by the excavations in Usu Area at Date city in Hokkaido, preliminary reports. *Proceedings of 29th Congress of Japan Society for Scientific Studies on Cultural Properties*, 70–71.

Sōmu-shō (2013) 2013 *Jōhō Tsūshin Hakusho* (2013 Information and Communication White Paper). Accessed on September 1, 2017 at http://www.soumu.go.jp/johotsusintokei/whitepaper/h25.html.

Sōmu-shō (2017) *2017 Jōhō Tsūshin Hakusho* (2017 Information and

Communication White Paper). Accessed on September 1, 2017 at http://www.soumu.go.jp/johotsusintokei/whitepaper/h29.html.

Stiglitz, J., A. Sen and J.-P. Fitoussi (2009) *Report of the Commission on the Measurement of Economic Performance and Social Progress.* Accessed on September 7, 2017 at http://ec.europa.eu/eurostat/documents/118025/118123/Fitoussi+Commission+report.

Stothers, R. and M. Rampino (1983) Volcanic eruptions in the Mediterranean before AD630 from written and archaeological sources. *Journal of Geophysical Research,* 88: 6357–6371.

Sugihara, T. (1996) *Kokusai Shihō Saiban Seido* (System of the International Court of Justice). Tokyo: Yūhikaku.

Sultan bin Salman bin Abdul-Aziz Al Saud (1985) Opening remarks at the First Congress of the Association of Space Explorers, Cernay, France (October 1985). Accessed on September 1, 2017 at http://www.space-explorers.org/congress/congress1.html.

Suzuki, M. (2007) *Heiwa to Anzen Hoshō Kokusai Kankeiron Series 2* (Peace and security international relations series 2). Tokyo: University of Tokyo Press.

Tachibanaki, T. (2006) *Kakusa Shakai: Nani ga Mondai nanoka* (Unequal society: What are the problems?). Tokyo: Iwanami Shoten.

Tani, Y. (2008) *Kinsei Yoshino Ringyōshi* (Early-modern history of Yoshino forestry). Kyoto: Shibunkaku.

Tashakkori, A. and C. Teddlie (eds) (2002) *Handbook of Mixed Methods in Social and Behavioral Research.* Thousand Oaks: SAGE Publications.

Thomson, R. and J.K. Pritchard et al. (2000) Recent common ancestry of human Y chromosomes: Evidence from DNA sequence data. *Proceedings of the Natural Academy of Sciences,* 97: 7360–7365.

Tomonaga, S. (1979) *Butsurigaku towa Nandarōka* (What is physics?). Tokyo: Iwanami Paperbacks.

Totman, C. (1998) *Nihonjin wa Donoyōni Mori o Tsukutte Kitanoka.* Japanese translation of *The Green Archipelago, Forestry in Preindustrial Japan,* trans. M. Kumazaki. Tokyo: Tsukiji Shokan.

Uemura, T. (2014) *Gurōbaru Kinyū ga Chikyū Kyōyū-Zai ni Naru tame ni 'Takkusu Heibun,' 'Gyanburu Keizai' ni Taisuru Shohōsen', 'Gurōbaru Komonzu to Kokuren* (Global finance becoming the world's common goods: Tax havens, 'prescription for the gambling economy'). Tokyo: Kokusai Shoin.

Umesao, T. (1957) *Bunmei no Seitaishikan* (Eco-historical view of civilization). Tokyo: Chuokoron Sha Publishers. English version published in 2003 as *An Ecological View of History: Japanese Civilization in the World Context.* Melbourne: Trans Pacific Press.

UNEP (2011) *Towards a Green Economy: Pathways to Sustainable Development and Poverty Eradication.* New York: United Nations.

UNESCAP (2012) *Low-Carbon Green Growth Roadmap for Asia and the Pacific.* Bangkok: United Nations Economic and Social Commission for Asia and the Pacific.

UNESCO (2015) *Education for All 2000-2015: Achievements and challenges (EFA Global Monitoring Report 2015).* Paris: UNESCO Publishing.

UNFCCC (2007) *Climate Change: Impacts, Vulnerabilities and Adaptation.* Bonn: Climate Change Secretariat (UNFCCC).

United States Department of Education (1998) *Reducing Class Size: What Do We Know?* Washington, DC: United States Department of Education.

United States Department of Health and Human Services, Administration for Children and Families (2010) *Head Start Impact Study: Final Report.* Washington, DC.

United States Government (1980) *The Global 2000 Report.* Washington DC: US Government Printing Office.

University of Illinois at Urbana-Champaign (2004) Shutdown of circulation pattern could be disastrous, researchers say. *ScienceDaily,* December 20. Accessed at www.sciencedaily.com/releases/2004/12/041219 153611.htm.

Usui, Y. (2006) Jizoku Kanōna Hatten to Kikō Hendō (Sustainable development and climate change. In S. Katsuhiro (ed.) *Kokusai Kikō* (International organizations). Tokyo: Iwanami Shoten.

Van de Mortel, E. (2002) *An Institutional Approach to the Transition Processes.* Aldershot: Ashgate.

Vellinga, M. and R.A. Wood (2008) Impacts of thermohaline circulation shutdown in the twenty-first century. *Climatic Change,* 91: 43–63.

Watts, D.J. and S.H. Strogatz (1998) Collective dynamics of 'small-world' networks. *Nature,* 393: 440–442.

WCED (1987) *Our Common Future.* Oxford: Oxford University Press.

Weinberg, A.M. (1972) Science and trans science. *Minerva,* 10: 209–222.

Weiss, C.H. (1979) The many meanings of research utilization. *Public Administration Review,* 39(5): 426–431.

Weiss, C.H. and M.J. Bucuvalas (1980) Truth tests and utility tests: Decision-makers' frames of reference for social science research. *American Sociological Review,* 302–313.

Weng, H., G. Wu, Y. Liu, S.K. Behera and T. Yamagata (2009) Anomalous summer climate in China influenced by the tropical Indo-Pacific Oceans. *Climate Dynamics,* 36: 769–782.

Weng, H., K. Ashok, S.K. Behera, S.A. Rao and T. Yamagata (2007) Impacts of recent El Nino Modoki on dry/wet conditions in the Pacific rim during boreal summer. *Climate Dynamics,* 29: 113–129.

Weng, H., S.K. Behera and T. Yamagata (2009) Anomalous winter climate conditions in the Pacific rim during recent El Nino Modoki and El Nino events. *Climate Dynamics,* 32: 663–674.

Widrow, B., R. Hartenstein and R. Hechtnielsen (2005) 1917 Karl Steinbuch 2005. *IEEE Computational Intelligence Society 5.* Accessed on September 1, 2017 at http://helios.informatik.uni-kl.de/euology.pdf.

Wiener, N. (1965 [1948]) *Cybernetics or Control and Communication in the Animal and the Machine.* Cambridge: MIT Press.

Wikipedia (n.d.-a) Big data. Accessed on September 1, 2017 at http://en.wikipedia.org/wiki/Big_data.

Wikipedia (n.d.-b) Hanshin Awaji Daishinsai (The Great Hanshin-Awaji Earthquake). Accessed on September 1, 2017 at http://ja.wikipedia.

org/wiki/%E9%98%AA%E7%A5%9E%E3%83%BB%E6%B7%A1
%E8%B7%AF%E5%A4%A7%E9%9C%87%E7%81%BD.

Wikipedia (n.d.-c) Ubiquitous computing. Accessed on September 1, 2017 at http://en.wikipedia.org/wiki/Ubiquitous_computing.

World Bank (2012) *Inclusive Green Growth: The Pathway to Sustainable Development.* Washington, DC: International Bank for Reconstruction and Development.

World Commission on Environment and Development (1987) *Chikyū no Mirai o Mamoru tameni* (Our common future). S. Ōkita (ed.). Tokyo: Fukutake.

Yamaguchi, E. (2005) *Inobēshon, Hakai to Kyōmei* (Innovation: Paradigm disruptions and fields of resonance). Tokyo: NTT Publishing Co., Ltd.

Yamaguchi, E. (2014) *Shinu made ni Manabitai Itsutsu no Butsurigaku* (Five physics theories to learn before you die). Tokyo: Chikuma Shobō.

Yamaguchi, E. and C. Miyazaki (2007) *Jēru Fukuchiyamasen Jiko no Honshitsu: Kigyō no Shakaiteki Sekinin o Kagaku kara Toraeru* (The essence of the JR Fukuchiyama train crash: Define social responsibilities of corporates from the perspective of science). Tokyo: NTT Publishing Co., Ltd.

Yamaguchi, E., Y. Nishimura and M. Kawaguchi (2012) *Fukushima Repōto: Genpatsu Jiko no Honshitsu* (FUKUSHIMA report: Nature of the nuclear accident). Tokyo: Nikkei BP consulting.

Yuan, C. and T. Yamagata (2014) California Niño/Niña. *Scientific Reports*, 4(4801).

Index